Encyclopedia of
Human Body Systems

Encyclopedia of Human Body Systems

VOLUME 2

Julie McDowell, Editor

AN IMPRINT OF ABC-CLIO, LLC
Santa Barbara, California • Denver, Colorado • Oxford, England

Library of Congress Cataloging-in-Publication Data

McDowell, Julie.
   Encyclopedia of human body systems / Julie McDowell.
      p. cm.
   Includes bibliographical references and index.
      ISBN 978–0–313–39175–0 (hard copy : alk. paper) — ISBN 978–0–313–39176–7 (ebook)
   1. Human physiology—Encyclopedias. I. Title.
QP11.M33   2011
612.003—dc22   2010021682

ISBN: 978–0–313–39175–0
EISBN: 978–0–313–39176–7

14 13 12 11 10   1 2 3 4 5

This book is also available on the World Wide Web as an eBook.
Visit www.abc-clio.com for details.

Greenwood
An Imprint of ABC-CLIO, LLC

ABC-CLIO, LLC
130 Cremona Drive, P.O. Box 1911
Santa Barbara, California 93116-1911

This book is printed on acid-free paper ∞

Manufactured in the United States of America

Contents

7

The Muscular System

Amy Adams

Interesting Facts

- The body contains roughly 630 skeletal muscles.

- The skeletal muscles account for roughly 50 percent of the body weight in men, 40 percent of the body weight in women, and 25 percent of a baby's body weight.

- After age 50, people lose about 10 percent of their muscle fibers per decade.

- Resting muscles receive about 20 percent of blood flow.

- During heavy exercise, the muscles receive from 60 to 85 percent of blood flow.

- Five to ten percent of a person's body weight is heart and smooth muscle.

- A fast-twitch muscle reaches peak contraction in about 1/20 of a second.

- A slow-twitch muscle reaches peak contraction in about 1/10 of a second.

- A single motor unit can range from two to three muscle fibers in the larynx to 2,000 fibers in the hamstring.

- Each sarcomere shortens between one-half to one-third its total length during a contraction.

- Muscles use up stored adenosine triphosphate (ATP) in about two seconds.

- Most pain that occurs the day after exercise is the result of eccentric exercise.

- Pain following exercise occurs most often in the region of the muscle farthest from the center of the body.

- Most strains happen at the junction between the muscle and the tendon.

- The second-most common lethal genetic disease for a child to be born with is muscular dystrophy.

- One in 3,300 male babies is born with Duchenne muscular dystrophy.

Chapter Highlights

- Types of muscles

- Muscle fibers

- Muscle contractions

- Role nerves play in muscle function

- Slide filament model

- Fast-twitch and slow-twitch muscles

- How muscles use energy

- Role of oxygen in muscle function

- Muscle's fuel sources

- Muscle behavior in anaerobic and aerobic exercise

- Changes to heart and smooth muscle during exercise

Words to Watch For

Acetyl coenzyme A

Acetylcholine

Actin

Aerobic exercise

Anaerobic exercise

Antagonistic muscle
   pairs

Calorie

Cardiac muscle

Citric acid cycle

Concentric
   contraction

Creatine phosphate

Eccentric contraction

Fascia

Fascicles

Fast-twitch muscle
   fibers

Glycogen

Glycosis

H zone

Hypertrophy

I band

Insertion

Insulin

Intercalated disk

Isometric contraction

Mitochondria

Muscle fibers

Myofibrils

Myoglobin

Myosin

Origin

Oxidative
   phosphorylation

Perimysium

Pyruvate

Sarcolemma

Sarcomeres

Sarcoplasmic
   reticulum

Skeletal muscle

Slow-twitch muscle
   fibers

Smooth muscle

Sphincters

Synergist

T tubules

Tendon

Testosterone

Tetanus contraction

Tropomyosin

Troponin

Z band

Introduction

One characteristic that unites all animals is the ability to move from place to place. Animals walk, crawl, or swim to find food, avoid prey, reproduce, and live out their lives. All of this movement requires muscles. In simple organisms, a few muscles are enough to control how the animal gets around. But in complex animals such as humans, a complex network of different muscle types helps fine-tune our movements. Large, powerful muscles allow us to walk, while delicate muscles give us the ability to make the detailed movements needed to write or play the piano. In addition to helping us navigate the world, muscles are essential to our internal processes. The heart beats more than one billion times in an average human life, and each of those beats is the result of the heart muscle contracting. Other muscles in the body are responsible for moving food through the digestive system, causing air to rush into and out of the lungs, or directing the blood as it flows through the circulatory system. This chapter provides a detailed overview of the anatomy, physiology, and medical issues affecting the many muscles in our body.

Humans have three distinctly different types of muscles: Our roughly 600 skeletal muscles allow us to navigate the world; the heart muscle keeps the heart beating; and smooth muscles line our internal organs, digestive tract, and veins. Although people have been interested in the question of how muscles contract since the era of ancient Greek civilization, not until the invention of the microscope in the mid-1800s did scientists first identify the three different types of muscles. Even then, scientists did not understand how the muscles contracted until 1952, when the newly invented electron microscope revealed the fine details of the muscles. Until that time, scientists had argued about how the proteins within the muscles interact to cause the muscle to contract. Another long-standing question had concerned what gives the muscles a signal to contract. As long ago as the 1500s, early scientists understood that nerves were needed to deliver signals to muscles, but without a detailed understanding of the nervous system and circulatory system, scientists studying the muscles could not fully understand how they function.

With all the functions they perform, muscles interact closely with other systems in the body. The blood vessels winding throughout muscles provide food and oxygen while carrying away waste products. Nerves deliver the essential signal that causes the muscles to contract and also play a role in shaping how muscles develop in the growing embryo. Some muscles are critical in order for other body systems to function normally. The digestive system, for example, relies on smooth muscles to push food from one end of the body to the other, and smooth muscles lining the bladder regulate urination. With their widespread functions, muscles are involved in just about every aspect of human physiology.

The body's muscular system is very complex, and this chapter provides an overview of some of the system's highlights, including the major muscle groups and the composition of muscles and its fibers. This chapter will also look at how muscles contract, and how muscles use energy from various sources.

Muscle Organization and Actions

Before discussing the specifics of how muscles function, it is important to get an overview of terminology associated with different muscles and their actions. This section will go over how the actions of muscles are described, as well as muscle names and their functions.

There are specific terms for the actions of the muscles. These actions are most often grouped in pairs, and describe antagonistic functions (with the exception of rotation), as depicted in Table 7.1.

There are different ways to group the body's muscles, but this chapter will describe them based on four groups: those in the head and neck, the trunk, the shoulder and arm, and the hip and leg. The posterior and anterior illustrations of some of the major muscles are depicted in Figures 7.1 and 7.2. Each of these groups is described in Table 7.2.

TABLE 7.1
Actions of the Muscles

Action	Description, example
Flexion	Decrease a joint's angle, such as when lifting a knee up
Extension	Increase the joint's angle, such as when lowering a knee that is lifted
Adduction	Moving closer to the body's middle, such as when moving the arm towards the waist
Abduction	Moving away from the midline, such as when moving the arm up above the shoulders
Pronation	Turning the palm of the hand up
Supination	Turning the palm of the hand down towards the ground
Dorsiflexion	Lifting or elevating the foot
Plantar flexion	Lowering the foot, such as when pointing one's toes
Rotation	Moving a muscle or bone around its longitudinal axis, such as rotating a shoulder

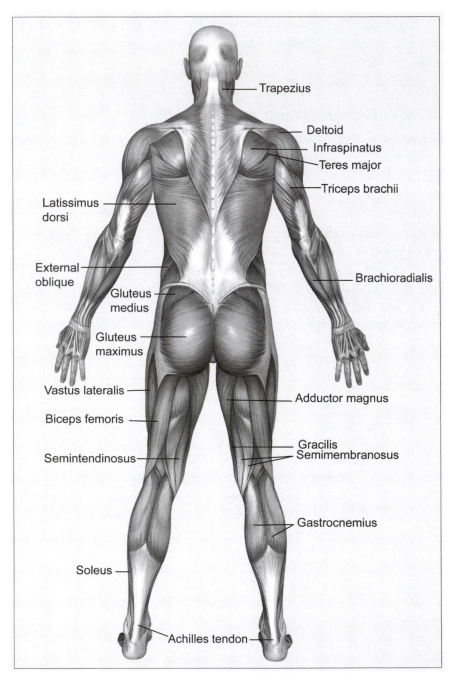

Figure 7.1   Posterior view of the body's muscular system. (Linda Bucklin/Dreamstime.com)

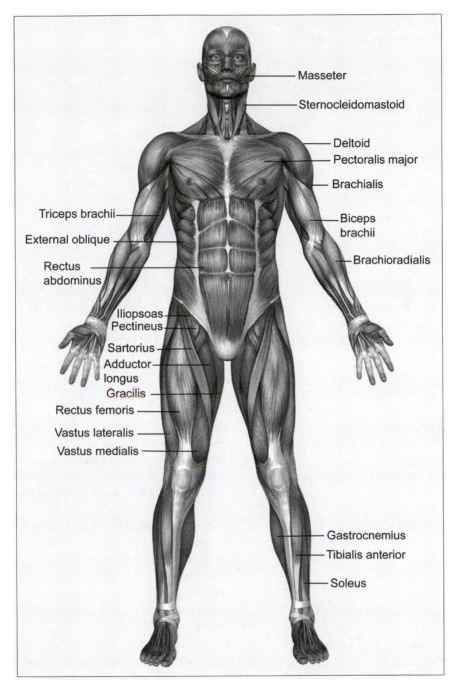

Figure 7.2   Anterior view of the body's muscular system. (Linda Bucklin/ Dreamstime.com)

TABLE 7.2
Major Muscle Groups

Name	Function
Major muscles of the head and neck	
Orbicularis oris	Gathers and puckers lips
Orbicularis oculi	Closes lid over eye
Masseter	Shuts jaw
Buccinator	Moves corners of mouth
Sternocleidomastoic	Flexes head and neck muscles to turn head towards opposite side
Semispinalis capitis	Extends head and neck muscles to turn head to the same side
Trunk muscles	
Trapezius	Raises and lowers shoulders
External intercostals	Allows inhalation by pulling ribs up and out
Internal intercostals	Forces exhalation by pulling ribs down and in
Diaphragm	Prepares chest cavity for inhalation by pulling down and flattening, which enlarges the cavity
Rectus abdominus	Constricts abdomen
Sacrospinalis muscles	Group of deep muscles that extends the vertebral column
Shoulder and arm muscle group	
Deltoid	Helps to abduct the muscles in the lower posterior neck, back of the lower shoulder region
Pectoralis major	Abducts the muscles in the upper anterior of the chest
Latissimus dorsi	Abducts and extend the muscles in the lower posterior of the back
Teres major	Abducts and extends the muscles in the upper posterior of the back
Triceps brachii	Extends the muscles of the forearm
Biceps brachii, brachioradialis	Extends and flexes the muscles of the forearm
Hip and leg muscle group	
Iliopsoas	Flexes the muscles of the femoris (femur), the long bone located in the thigh

TABLE 7.2 (*Continued*)

Name	Function
Gluteus maximus	Extends the muscles of the femur or thigh
Gluteus medius	Abducts the femur muscles
Quadriceps femoris group (rectus femoris, vastus lateralis, vastus medialis, vastus intermedius)	Flexes femur muscles and extends muscles of the lower leg
Hamstring muscle group (biceps femoris, semimembranosus, semitendinosus)	Extends femur muscles and flexes muscles of the lower leg
Adductor muscle group	Adducts femur muscles
Sartorius	Flexes lower leg muscles and femur
Gastrocnemius, Soleus	Plantar flexion of the foot (lowering movement)
Tibialis anterior	Dorsiflexion of the foot (elevating movement)

Anatomy of the Muscular System

Three basic types of muscles carry out all movement within the body: **smooth muscle**, **cardiac (heart) muscle**, and **skeletal muscle**. Although each of these three types of muscles shares the ability to contract, they are located in different places in the body, contract at different strengths, and look different under a microscope.

Types of Muscles

Smooth Muscle

Smooth muscle lines many of the internal organs. They cause contractions that move food through the intestines, expand and contract the blood vessels to regulate blood supply, and contract to push the baby out of a woman's uterus. Smooth muscles also control the size of the pupil and are found at the base of hair follicles, where they produce goose bumps when it is cold. Ideally, these goose bumps would raise the hair to trap an insulating layer of air around the body, although few humans have enough hair for goose bumps to serve a practical purpose.

Smooth muscles are also called involuntary muscles because they cannot be controlled voluntarily; one cannot decide to relax the pupil to let in more light any more than a person can stop the spread of a blush across the face. Instead, smooth muscles contract in response to biological conditions such as food passing through the intestine, bright light shining on the eye, cold temperatures, or embarrassment.

Smooth muscles contract slowly but with great force, and can hold a contraction without growing tired. They also shorten more when they contract than other muscles do. Whereas most muscles are designed for quick, precise motions, smooth muscles are designed for long-term squeezing.

Smooth muscles earned their name by being smooth in appearance under the microscope. The individual cells are long and tapered and tend to form into sheets, such as in the lining of the digestive system. As with most cells in the body, each smooth muscle cell has its own nucleus.

Cardiac Muscle

Cardiac muscle is the muscle that makes up the heart. It is similar to smooth muscle in that (1) each cardiac muscle cell has its own nucleus and (2) the heart muscle cannot be controlled voluntarily. Instead, a region of the brain monitors how much oxygen is in the blood and adjusts how quickly the heart beats accordingly. A runner cannot decide to slow the heart rate, nor can a resting person voluntarily speed up how quickly the heart beats.

The cardiac muscle is extremely strong and is unique in that the entire muscle contracts at the same time. Whereas smooth muscles squeeze consistently, the cardiac muscle is unable to sustain a contraction. Instead, the entire muscle contracts forcefully, then relaxes. The cardiac muscle also looks different under the microscope than smooth muscle. Each individual muscle cell is cylindrical in shape and can have many branches. The cells join together to form long, branched tubes, with each cell separated by a disk of cell membrane called the intercalated disk.

Skeletal Muscle

The skeletal muscles are all the muscles that attach to the skeleton and help the body move. These are the 600 or so muscles that you might exercise at the gym and that you use to move around or pick up a book.

They are also the muscles that get strained from exercise and that become diseased in muscular dystrophy. Most of the muscles in the body are skeletal muscles. For that reason, throughout this chapter, the term muscles will refer specifically to skeletal muscles unless stated otherwise.

Skeletal muscles generally have both ends anchored to the skeleton with a thick, rope-like tissue called a **tendon**. When the muscle contracts, it pulls on the tendons, which then pull two skeletal regions closer together. Usually, two ends of the muscles are on either side of a joint. For example, the biceps muscle on the inside of the upper arm is attached to the shoulder at one end and to the forearm at the other end. The muscles on the inside of the joint, such as the biceps, pull two sides of the joint together, causing the joint to bend, while muscles on the outside of a joint contract to pull the joint straight.

Most muscles are organized into pairs located on either side of a joint. These two muscles work against each other and are thus called **antagonistic pairs**. One antagonistic pair might be the biceps muscle that bends the arm and the triceps muscle that straightens it back out. A similar pair is the quadriceps muscle on the front of the thigh that straightens the leg, and the hamstring on the back of the thigh that bends it. In each case, the two muscles have opposing actions and help keep the joint stable and control movement. Another type of muscle helps fine-tune direction of the movement caused by the antagonistic pair. This type of muscle works in synergy with a larger, stronger muscle and is called a **synergist**. For example, the strong bicep muscle bends the arm at the elbow, but synergist muscles control whether the hand moves directly toward the shoulder, or veers right or left.

To understand how agonists and antagonists work, imagine lifting a weight. As one lifts the weight, the biceps contracts, causing the arm to bend, and the triceps relaxes to allow the bend. The triceps does stay somewhat contracted to help control how quickly the arm moves. When one lowers the weight, the triceps contracts to pull the arm straight and the biceps slowly relaxes, allowing the weight to lower (Sidebar 7.1).

Skeletal muscle has things in common with both smooth and cardiac muscle, but is unique in that it is the only muscle group that a person can move voluntarily. Like the cardiac muscle, each fiber can only contract once in response to a signal from a nerve. This single contraction is

---

SIDEBAR 7.1

Therapy for Your Muscles?

Many people rely on massage therapy to ease pain and soreness related to sports injuries, as well as to increase relaxion and reduce stress, anxiety, and depression. In fact, an estimated 18 million adults in the United States are believed to receive massage therapy every year. People use massage for a variety of health-related purposes, such as to relieve pain or rehabilitate sports injuries, as well as to reduce stress, increase relaxation, address anxiety and depression, and aid general wellness. However, the National Institutes of Health's National Center for Complementary and Alternative Medicine emphasizes that there is limited scientific evidence on massage therapy; therefore, it is not clear what impact massage therapy has on muscles, if it impacts health, and if so, how it influences health.

There are various types of massage. Some examples include Swedish, deep tissue, and trigger point massage. In a Swedish massage, the masseuse uses long stroke and deep circular movements, as well as kneading, vibration, and tapping on the muscles. Deep tissue and trigger massage focuses on knots in the muscles that can become painful when pressure is applied, although the massaging works to ease these knots.

---

called a simple twitch. However, most movements like walking or playing basketball involve smooth, sustained motion that could hardly be achieved by simple twitches. Instead, signals from the nerves come in extremely quick pulses. Rather than resulting in a series of simple twitches, these signals cause a sustained contraction called a **tetanus contraction**. (The disease tetanus—which is also called lockjaw—causes muscles of the body to contract).

Anatomy of a Skeletal Muscle

Most muscles connect to a bone at either end. The tendons that attach muscles to bones are also part of the bone's outer coating and part of the muscle's coating, making the connection extremely strong. One end of the muscle is generally considered to be the **origin**, while the other end is the **insertion**. The origin of the muscle is usually closer to the center of the body and is connected to a bone that does not move much when the muscle contracts. The insertion is usually the end of the muscle that is farther from the middle of the body and is connected to the bone that moves when the muscle contracts. For example, when the biceps muscle flexes, the forearm moves closer to the shoulder. With that in mind, the end that connects to the shoulder is the origin, while the end that connects to the forearm is the insertion.

Although most muscles are attached at both ends to a bone, a few muscles are not. These include the facial muscles that allow a person to smile or frown and the many muscles that make up the tongue. These muscles have complex origins and insertions that allow the tongue to do everything from pick food out of the teeth to control whistling and speech. Another group of muscles, such as those in the abdomen, connect to bands of tendons that cross the stomach rather than to any bone. When the muscles contract, they pull those tendons toward each other to bend the stomach.

There are also circular bands of muscles called **sphincters** that help control the flow of food through the digestive system. These sphincters do not connect to muscles, tendons, or skin, but instead form a continuous, circular band of muscles that surround the mouth, the ends of the stomach, and the anus. Puckering the lips for a kiss involves contracting a sphincter muscle. These muscles are especially good at constricting passages, such as at either end of the stomach where they prevent food from leaving until it is fully digested.

Layers of the Muscle

Although a muscle looks like a single, solid mass, it is actually made up of many smaller fibers all bundled together to form one functioning unit. The entire muscle is wrapped in a connective tissue called the **fascia**, or the epimysium, that holds the muscle together. It is this fascia that forms into tendons and is fused to the bone to make a strong muscle-bone connection. Held within this outer covering are many smaller bundles of fibers called the **fascicles**. These bundles are also held together with connective tissue called the **perimysium**. These fascicles cause the stringiness one may notice when eating meat (see Figure 7.3 for an illustration of muscle anatomy).

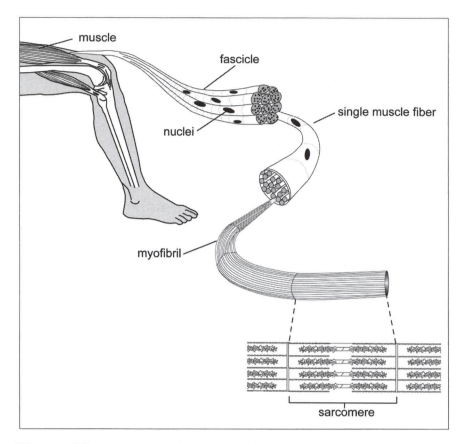

Figure 7.3   Anatomy of a muscle. (Sandy Windelspecht/Ricochet Productions)

Within the fascicles are the **muscle fibers** themselves. These fibers are actually very long cells—each less than the width of a hair—with many nuclei. Most cells only have one nucleus. Muscle cells are a conglomeration of many cells that fuse during development into one long fiber with several nuclei per fiber. The longest single-muscle fibers in humans tend to be about 4.7 inches (12 centimeters). The number of individual fibers in each muscle can range from 10,000 fibers to more than a million fibers, with each fiber spanning only a portion of the muscle's full length in some long muscles.

Finally, each muscle fiber contains thousands of long units called **myofibrils**. These myofibrils are what actually contract when the cell receives a contraction signal from a nerve.

Running through the muscle between fascicles are nerves and blood vessels. The nerves relay signals that run from the brain, down the spinal column, and through nerve cells to the muscle. When a person decides to move, these nerves relay that signal to the muscle fibers that carry out that decision. The blood vessels deliver food, nutrients, and oxygen to the muscles to help the muscles contract. They also pick up carbon dioxide from the muscle and deliver it to the lungs where it is breathed out in exchange for more oxygen.

Muscle Fiber Anatomy

The individual muscle fibers are surrounded by a cell membrane called the **sarcolemma**. The sarcolemma both encircles the muscle fiber and sends hollow projections across the fiber. These channels, called **T tubules**, help the muscle transmit the signal to contract.

Just underneath the sarcolemma are granules of a type of sugar called **glycogen**. This glycogen serves as a food reservoir for the muscle to feed on when it is in heavy use. Along with the glycogen, cellular units called **mitochondria** also inhabit the space just under the sarcolemma. These mitochondria are referred to as the "powerhouse of the cell." They convert food from the bloodstream, glycogen, and other sources into energy that the muscle uses in order to contract.

Under the mitochondria is a network of hollow tubules called the **sarcoplasmic reticulum**. The sarcoplasmic reticulum is a repository for calcium, which the muscle uses in the process of contraction. It closely follows the T tubules that cross the fiber and winds throughout the muscle fiber, providing a quick source of calcium.

Ultrastructure of a Muscle Fiber

Skeletal muscle is also called striated muscle, because under a microscope, the myofibrils contain many tiny stripes, or striations. The stripes that can be seen under the microscope are the individual units of the

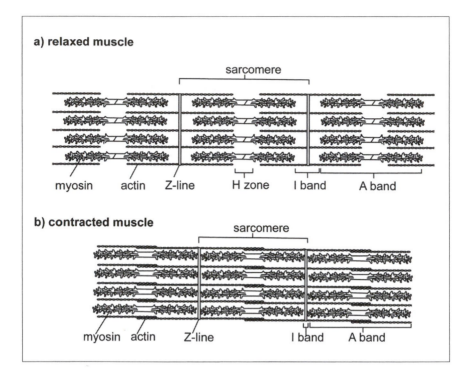

Figure 7.4 Sarcomere Structure. The structure, when relaxed and contracted. (Sandy Windelspecht/Ricochet Productions)

muscle fiber, which are called **sarcomeres** (Figure 7.4). Cardiac muscle also has these striations, though smooth muscle does not.

It is bounded on both ends by a vertical band of tissue called a **Z band**. Looking closely at the sarcomere under a microscope, there are several horizontal stripes of protein filaments. These are made up of the two primary types of protein filaments in a muscle: **actin** and **myosin**. There are roughly 1,500 myosin filaments and 3,000 actin filaments per muscle fiber.

Emanating from both Z bands are the long, thin filaments of actin. These stick out into the center of the sarcomere like posts, with one end embedded in the Z band. When a muscle is at rest, these two sets of actin filaments do not overlap, leaving a gap in the center of the sarcomere.

Myosin filaments lie parallel to the actin in the center of the sarcomere and overlap with the actin on both ends. These myosin filaments are tethered in place to either Z band by a fine, thread-like filament called **titin**.

The bands of actin and myosin within the sarcomere account for the striations that can be seen under the microscope. The region next to the Z band that contains only actin filaments is called the **I band**. The region between actin filaments in the center of the sarcomere is called the **H zone**. The entire span of the myosin is called the **A band**.

The actin filaments are actually made up of three proteins. The globular actin proteins form into two chains that wrap around each other like two strands of twisted pearls. Winding around the actin are long, narrow filaments of a protein called **tropomyosin**. Tropomyosin fits into the groove between the two actin strands. At intervals, a smaller protein called **troponin** dots the outside of the actin/tropomyosin complex.

Myosin is much thicker than actin and thus is sometimes referred to as the "thick filament." It looks like a chain of golf clubs twisted around each other. Each golf club represents a single myosin molecule that doubles back on itself at the tip to form the head.

Anatomy of a Contraction

Skeletal Muscle

Each contraction results when a nerve delivers a message to the muscle. The smooth and cardiac muscles respond to nerves that are controlled by automatic reactions in the brain. Skeletal muscles move in response to nerves controlled by a person's decisions.

When a muscle gets a signal to contract, the globular myosin heads attach to the actin and pull, much like a team of people playing tug-of-war. The heads flex, release, and reattach farther up the actin filament, then flex again. This action works like a ratchet pulling the myosin along the actin filament. Because the actin filaments are attached to the Z bands at both ends of the sarcomere, this action pulls the actin filaments toward each other and pulls the Z bands closer together to shorten the sarcomere.

This mechanism is called the sliding filament model for muscle contraction. One result of the filaments sliding past each other is that the actin filaments overlap in the center of the sarcomere, causing the sarcomere to grow thicker. This overlap is what accounts for the muscle's bulge when flexed. Although each individual sarcomere contracts only a small amount, when many sarcomeres contract at once, the muscle gets considerably shorter.

When a sarcomere contracts, the A band (or myosin) remains the same size in the center of the sarcomere because the myosin does not change length. The H band, which is the gap between the actin filaments in the center of the sarcomere, disappears as the actin filaments are pulled toward the center of the sarcomere. Likewise, the I band disappears during a contraction because the space between the Z band and the myosin filaments grows shorter. Notice that as the sarcomere contracts, the filaments stay the same length, although the distance between the Z bands decreases.

The strength of a contraction depends in part on how many muscle fibers (and their associated myofibrils) receive the signal to contract. When only a few fibers receive a signal, the contraction is relatively weak. When many fibers receive a signal, many more sarcomeres will contract, making for a much stronger contraction.

Cardiac Muscle

Cardiac muscle has actin and myosin arranged into striated sarcomeres, much like skeletal muscle, and it has a well-developed sarcoplasmic reticulum. Unlike skeletal muscle, which contains single long fibers, cardiac muscle fibers have many branch points and appear like a complex web. When this muscle contracts, the irregular shape of the branching fibers causes the heart muscle to twist and essentially wring blood out of the heart.

Smooth Muscle

Smooth muscles lack the orderly striations of skeletal and cardiac muscles. Instead, the actin filaments are arranged in a roughly parallel way and are attached to the ends of the cell. Myosin filaments slide the actin past each other and pull the ends of the cell closer together. In smooth muscle, this action causes the entire muscle sheet to contract.

Unlike either skeletal or heart muscle, smooth muscle does not have a well-developed sarcoplasmic reticulum. When a signal from a nerve arrives at the muscle, calcium seeps in from outside the cell. This process is much slower than in the other types of muscles and does not allow as much calcium into the cell. Overall, the contraction takes longer to start and does not pull as hard as the other muscle types.

Contraction Strength

When a nerve sends to a skeletal muscle fiber the signal to contract, that fiber contracts with an all-or-none response. It cannot contract partway. This seems to contradict day-to-day experience, in which the strength of a contraction can be adjusted for the relatively little strength required to pick up a pencil or the large amount of strength needed to pick up a large weight. If each contraction used the muscle's complete strength, then it would be nearly impossible to pick up a pencil in a controlled manner.

It turns out that the amount the muscle contracts has to do with how many fibers receive a signal. One single muscle fiber cannot produce a contraction that is strong enough to do any significant work. For this reason, most single nerves have connections with about 150 fibers, each of which contracts when the nerve sends a signal. These groups of muscle fibers are called motor units. In areas like the hands, where fine control is needed for activities such as writing or playing a musical instrument, the motor units are composed of fewer fibers. Each nerve controls only a few fibers, providing the ability to make miniscule changes in how the muscle contracts. In muscles such as the hamstring, which provides the strength to bend the leg, the motor unit may be composed of a much larger number of fibers.

Stimulation from nerves also controls how long a contraction lasts. When a doctor taps the knee to test a person's reflexes, the leg twitches but does not stay contracted. That is because the muscle receives only a brief stimulation from the nerve. In order to keep the muscle contracted, the nerve sends a rapid volley of signals in quick succession. Although each signal stimulates only a single contraction, they add up so that the muscle can stay flexed to hold a paintbrush steady or hold a dance pose.

Some muscles, such as those that help a person stand upright, always have some muscle fibers contracting in order to maintain posture. In order to prevent one group of muscle fibers from getting tired, the brain sends signals to different motor units so that the groups of fibers share responsibility for holding a person upright. This same rotation takes place if one holds a heavy object for a long time. At first, a few groups of muscle fibers will all get the signal to contract. But over time, if those fibers become tired, the brain recruits a new group of fibers to take over responsibility for contracting. After a long time of standing upright or holding a heavy book, the muscle will become tired, but no one group of muscle fibers is damaged because the motor units shared the load.

One final factor that controls the strength of a contraction is the width of a single muscle fiber. Anyone who has been to a gym knows that the more a muscle gets used, the larger it is. This change has to do, in part, with the size of the individual muscle fiber, and therefore how many myofibrils are within the muscle cell. A thick muscle fiber with many myofibrils will respond with more strength in response to a signal from a nerve. In essence, that muscle cell has more myosin heads playing tug-of-war on the actin filaments, allowing that fiber to pull harder. Although the nerve will contact the same number of muscle fibers in a motor unit, the overall contraction will be stronger.

The Nerve/Muscle Connection

Skeletal Muscle

The signal to contract a muscle fiber comes from nerves. Each nerve starts at the spinal column, travels through the body, and then branches out so that the single nerve controls several muscle fibers in a motor unit. When the nerve is stimulated by the brain, an electrical signal travels the length of the nerve and reaches each of the many branched tips. At the tip, the electrical signal causes the nerve to release a chemical called **acetylcholine** into a gap between the nerve cell and the muscle fiber. The acetylcholine is also called a neurotransmitter because it transmits a signal from the nerve to the muscle.

Acetylcholine travels across the space between the nerve and muscle and binds to a protein (called a receptor) located on the muscle cell membrane. The bound acetylcholine triggers a series of reactions to occur that propagate the electrical signal down the muscle fiber in a wave, transmitting the signal throughout the muscle fiber and across the

T tubules almost instantaneously. Where the T tubules and sarcoplasmic reticulum come in close contact, the signal from the T tubules transfers to the sarcoplasmic reticulum. In response, the sarcoplasmic reticulum releases calcium into the muscle fiber. Keep in mind that because the nerve connects to many different muscle fibers, this same reaction takes place in each fiber at exactly the same time.

Heart Muscle

The same general principles hold true between skeletal muscle and heart muscle; however, there are some notable differences. In skeletal muscle, groups of fibers within the muscle contract as they are needed—the entire muscle rarely contracts at once. In the heart, on the other hand, the entire muscle contracts with each heartbeat. For this reason, the heart muscle does not have the web of nerves branching out to connect with each individual muscle fiber.

A signal to contract reaches the heart at a specialized group of muscle cells located near the top back side of the heart. This group of cells is called the pacemaker. When a nerve delivers its neurotransmitter to the pacemaker region, these cells spread the signal across muscle cells in the top half of the heart. Unlike skeletal muscle fibers, where the signal stays within the single fiber, heart muscle fibers are connected with an **intercalated disk** that allows a signal to move easily from one muscle fiber to the next. Because the signal can spread so quickly, all of the fibers in this top region contract at the same time. This first phase of the heartbeat pushes the blood down into the lower portion of the heart, while the upper region refills with blood.

When the pacemaker receives a signal to contract, it delivers the message to all the upper muscles and also sends the signal to a group of cells in the lower portion of the heart. This message is delayed slightly to give the upper muscles time to contract. When the lower region receives the message to contract, it spreads the signal to all the muscle fibers of the lower part of the heart. This contraction finishes the heartbeat and pushes blood out to the body. Chapter 2 on the circulatory system has more information about how the heart relays the signal to contract.

Unlike the skeletal muscle, heart muscles can beat without a signal from the brain and can continue beating even outside of the body.

Although the pacemaker cells generally relay a signal from a set of nerves, if the heart is disconnected from those nerves or if the nerves fail to fire, the pacemaker will continue sending a signal to contract at a regular rhythm. Mechanical pacemakers can back up the heart's pacemaker to ensure that the heart continues to beat at a regular rhythm, even if the heart's natural pacemaker fails to maintain the heartbeat.

Sliding Filament Model of Muscle Contraction

Skeletal and Heart Muscle

Researchers knew as early as 1883 that calcium was required in order for a muscle to contract. However, not until the 1960s did researchers understand what role calcium played when it was released from the sarcoplasmic reticulum.

It turns out that when either the skeletal or heart muscle is relaxed, the globular myosin heads cannot attach to the actin filaments. The troponin/tropomyosin complex blocks access to areas of the actin filament where the myosin head would normally bind. This physical block ensures that the muscle stays relaxed when there is no nerve signal. To prevent the muscle from contracting inappropriately, the muscle fibers actively move calcium from the cell into the sarcoplasmic reticulum, which winds throughout the cell.

When a nerve signal sweeps across the muscle fiber and through the T tubules, it is translated to the sacroplasmic reticulum, which responds by releasing the stored calcium. Once in the muscle fiber, calcium binds to the troponin and causes the troponin to change shape. In this new shape, the troponin/tropomyosin complex shifts and reveals the site on the actin filament where myosin binds.

Although the myosin can now bind to actin, it takes energy in order for the actin and myosin to pull past each other. This energy comes from a molecule called **adenosine triphosphate (ATP)** that is normally bound to myosin when the muscle is in a relaxed state. ATP is the form of energy that is produced primarily by the mitochondria. Although ATP itself can do no work, other proteins within the cell can break the ATP into a related molecule called **adenosine diphosphate (ADP)**, and in the process release some energy. ATP is like a match: it does not release any energy when it sits in a box, but if someone strikes the match, it releases enough energy to burn fingers, melt wax, or heat a small drop of water. When the match is burned out, like ATP, it can no longer be used.

In a relaxed muscle, myosin is bound to a molecule of ATP. When calcium enters the cell and troponin changes configuration, the myosin breaks the ATP into two units and in the process releases a small amount of energy. Myosin uses that energy to change shape and stretch out far up the actin molecule. The myosin then binds to the newly revealed binding site on actin and releases the ATP. Without ATP, the myosin relaxes back into a less energetic configuration while still bound to actin. This change in shape pulls the actin past the myosin. A new molecule of ATP then binds myosin, allowing myosin to break its bond with actin and resume its stretched-out shape, reaching farther up the actin filament (Figure 7.5).

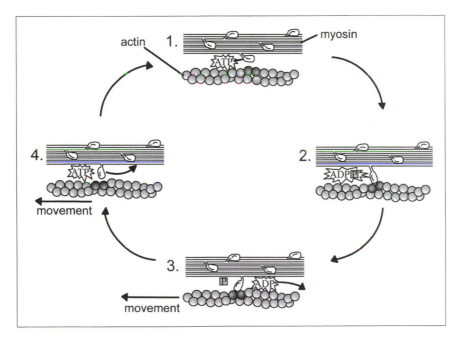

Figure 7.5 Myosin-Actin Filament. Myosin slides along the actin filament using ATP. (Sandy Windelspecht/Ricochet Productions)

The entire process repeats itself when myosin reattaches to the actin, relaxes, pulls the actin farther along, and binds a new molecule of ATP. With many myosin heads all attaching and flexing at different times, the process results in a slow, controlled contraction that lasts as long as ATP and calcium are present in the cell.

When a nerve stops firing, calcium stops being released from the sarcoplasmic reticulum, and calcium that is in the muscle fiber is actively transported back into storage. The troponin/tropomycin/actin complex then reverts back to a shape that prohibits myosin attachment. The myosin then cannot reattach, and the contraction ends. The actin filaments slide past the myosin and come back to rest with the Z bands farther apart.

One side effect of churning through ATP when a muscle contracts is that the process produces heat. People generally get hot when they exercise or use their muscles heavily; that is because each time myosin breaks ATP, a small amount of heat is released. When a person stops exercising and myosin no longer requires ATP, they eventually cool off. The body takes advantage of this phenomenon by shivering when a person gets cold. Using the muscles in this way acts as an internal heater to help warm the body. However, even with active shivering, the body cannot make enough heat to warm a person when he or she is very cold. Shivering is not considered to be a useful response, because it requires quite a bit of energy to shiver with very little actual heat being produced.

Another example of the contraction process comes when an organism dies. After death, the body no longer makes ATP. Remember that ATP is needed in order for myosin to release from the actin filament after it relaxes. Once a muscle has used up its store of ATP, the myosin heads can no longer release from the actin, locking the muscles into a rigid state that is also known as rigor mortis. In this condition, all of the muscles are tightly contracted, and the body is very difficult to bend.

Smooth Muscle

Smooth muscle does not contain the regular sarcomeres of skeletal and heart muscles. Instead, the actin filaments span the length of the entire muscle cell. Myosin pulls on those actin filaments to pull the cell shorter. This alternate form of contraction is also regulated in a different way.

In smooth muscle, the actin filament does not have a troponin/tropomysin complex preventing myosin from binding. Instead, calcium enters the cell and activates a set of enzymes. Those enzymes add a phosphate to myosin, changing its shape and allowing it to bind actin. These intermediary steps slow the time it takes for a smooth muscle to contract. Whereas heart and skeletal muscles contract immediately when calcium enters the muscle fiber, smooth muscle cells can take as long as a second to contract after calcium levels go up in the cell.

Types of Muscle Fibers

Fast-Twitch and Slow-Twitch Fibers

All skeletal muscle fibers have the same basic structure, but they do vary in subtle ways that can dramatically affect the performance of the muscle. Muscles contain two general types of fibers: **slow-twitch** (type I) and **fast-twitch** (type II). As their names imply, slow-twitch fibers contract slowly whereas fast-twitch muscles contract quickly after they receive a signal from a nerve. On average, slow-twitch fibers take about one-tenth of a second to reach their peak contraction, while fast-twitch muscles take about half that time.

The difference between slow-twitch and fast-twitch muscle fibers lies in how quickly the myosin can cycle through ATP. The more quickly the myosin can use the ATP, the more quickly it can finish one stroke and attach to myosin for another pull.

It turns out that myosin itself is basically the same in both types of fibers. The difference between the myosins lies in their ability to break ATP—an enzyme function called an ATPase. Slow-twitch and fast-twitch fibers have myosins with different ATPase activity. As one would imagine, slow-twitch fiber myosins have an ATPase that breaks ATP slowly, while fast-twitch fiber myosins have an ATPase that breaks ATP quickly.

The two types of muscle fibers also differ in how quickly the contraction begins. Fast-twitch fibers have a much more extensive sarcoplasmic reticulum network than slow-twitch fibers, allowing these fibers to receive calcium more quickly after a nerve signal than slow-twitch muscles. In the time it takes a fast-twitch muscle to flood with calcium and begin a contraction, a slow-twitch muscle is still only slowly filling with calcium, and few troponin molecules have changed shape.

Sometimes the difference between fast- and slow-twitch fibers is not absolute. Fast-twitch fibers can be classified as type IIa or type IIb. The type IIa fibers have fast myosin ATPase and can contract faster than slow-twitch fibers, but they also have more endurance than fast-twitch fibers. The type IIb fibers are the pure fast-twitch fibers.

On average, fast-twitch fibers begin contracting five to six times faster than slow-twitch fibers. With their quick response and fast contractions, it is no surprise that sprinters have a higher percentage of fast-twitch muscle fibers than distance athletes

Roles for Slow-Twitch and Fast-Twitch Muscles

Slow-twitch and fast-twitch fibers play distinctly different roles in a muscle. In an average muscle, slow-twitch fibers make up about half the muscle, while fast-twitch fibers make up the other half. The photo shows a muscle section composed of both slow-twitch fibers (stained dark) and fast-twitch fibers. Keep in mind that people differ dramatically in their muscle composition and that fiber composition can differ even between muscles in a single person. For example, the solius muscle, which is deep inside the calf, is made up of almost exclusively slow-twitch fibers in everyone. Although the fiber composition does vary among individuals, a person who has predominantly fast- or slow-twitch muscles in the legs will have a similar composition in the arms and other muscles. This means that a person who is a particularly good sprinter when running is likely to be a sprinter rather than a distance athlete at swimming and biking as well.

When a muscle first starts being used—for example, when a person begins walking—the slow-twitch muscles are the first to be called upon to contract. When these muscles fatigue, or when the contraction needs to be more powerful, the muscle will recruit a subset of the fast-twitch type Ia fibers. Only when both of these types of fibers grow tired or when great strength is needed does the muscle call upon the fast-twitch type IIb fibers.

In order to control which type of muscle fiber contracts, each nerve connects to either all fast-twitch or all slow-twitch muscle fibers.

The nerves themselves also differ. Nerves that connect to slow-twitch fibers are smaller and connect to about 10 to 180 fibers in a motor unit. A nerve for fast-twitch fibers is much larger and connects to as many as 300 to 800 fibers in a motor unit. With this arrangement, the muscle can recruit a very small number of slow-twitch muscles at a time, but can signal many more fast-twitch muscle fibers with a single nerve impulse. Because of this, a single fast-twitch motor unit contraction is much stronger than the contraction of a single slow twitch motor unit.

Types of Contractions

In general, when a muscle receives a signal to contract, the muscle grows shorter and causes a joint to either straighten or bend. Although this basic principle holds true, there are other types of muscle contraction. In most activities such as running and jumping, these different types of contraction work smoothly together to create the motion. However, it is worth discussing these actions separately to point out different mechanisms of moving the muscle.

Concentric Action

A **concentric contraction** is the motion that a person generally associates with muscle contraction. In a concentric action, the muscle contracts and pulls two bones closer together, causing the joint to open or close. Concentric action would occur if one holds a weight in the hand and bends the elbow to flex the biceps.

The myosin heads can be thought of as a team of tug-of-war players and actin as their rope. Concentric action would occur when the team pulls and the rope moves toward them.

Isometric (Static) Action

Although people usually think about contracting a muscle to move a joint, there are other types of contractions. Imagine holding a weight that is too heavy to lift. Flexing the biceps muscle, the arm will remain straight even though the muscle is trying to contract. This is an **isometric contraction**. In an isometric contraction, the nerve sends a signal for the muscle to contract.

The muscle fiber floods with calcium, and the myosin pulls on the actin filaments, but the fiber does not generate enough force to bend the arm. This type of contraction also occurs when holding a weight in a stationary position. The muscle is flexing even though the weight is not moving. Only enough muscle fibers are activated to hold the weight steady. Pushing on a wall is also an isometric action.

In the same example, if a person grows tired of holding the weight and decides to lift the weight to a new position, the muscle will recruit extra muscle fibers to help lift the load. These extra fibers then provide enough strength to lift the weight, and the isometric action becomes a concentric action. Isometric action has to do with the weight of the object and the number of muscle fibers recruited to do the work. Going back to that same team of tug-of-war players, an isometric action would occur when the players pull with all their strength, but the rope does not budge.

The word isometric is also used to describe a type of exercise. Isometric exercises involve flexing a muscle without moving the joint. Isometric exercises were made popular in the 1950s as part of the Charles Atlas program. Although people did become stronger using this program, their muscles were particularly stronger only when they were bent at the same angle as was used in the isometric exercises. Although isometric exercise is far from perfect, it has the advantage of requiring little equipment or room. For this reason, it is used by astronauts trying to maintain muscle strength in the weightlessness of space.

Eccentric Action

After bending the arm to lift a heavy weight (a concentric action), the arm must then straighten to let the weight down. Straightening the arm requires the biceps to maintain some contraction to control how quickly the arm straightens. This contraction that occurs as the muscle lengthens is called an **eccentric contraction**. Even though the myosin heads are pulling on the actin filaments, those actin filaments are moving farther away from each other within the sarcomere. In tug-of-war terms, an eccentric action would occur when the players pull with all their strength, but the rope moves away from them and toward the other team.

Eccentric actions tend to damage the muscle fibers. Although this sounds like a bad thing, it can actually have positive results. After a session of exercise with eccentric action, the muscle fibers will become damaged and very sore. But they will adapt to just one session of eccentric exercise by increasing the number of sarcomeres per muscle fiber (essentially making each sarcomere shorter). With more sarcomeres per fiber, each sarcomere has to stretch less for the eccentric action, so the muscle sustains less damage. The other advantage to this is that the muscle will become quicker to respond to a signal because many more sarcomeres are doing the same work. This change lasts about 10 weeks—after that time, the muscle loses the adaptation and will once again be damaged by eccentric action.

Most people have experienced pain associated with eccentric action when they run or walk downhill. When a person walks downhill, the thigh muscle contracts as the leg extends in front. Transferring weight onto the other leg, the knee lowers slightly to absorb the shock and in the process stretches the thigh muscle. Stretching the thigh muscle while the muscle contracts is a form of eccentric action that damages the muscle and can lead to pain and weakness over the next two to three days. Walking or running downhill in the next few weeks will cause much less pain in the following days because the muscles have adjusted to the eccentric action.

Energy Use by Muscles

Muscles require energy in the form of adenosine triphosphate (ATP) in order to contract. This ATP is used as an energy source for all the reactions that take place in the body—it is required to conduct signals along the nerves, to translate genes into proteins, to move molecules into and out of a cell, and to contract muscles, among other things.

Each cell is self-sufficient when it comes to making ATP. A person cannot eat ATP and have the molecule transported to cells, nor can energy-starved cells take ATP from the bloodstream. Instead, each cell generates its own ATP from sugar, fat, or protein that comes either from the bloodstream or from the cell's internal stores. For this reason, cells that require large amounts of energy, such as muscle cells, have to be extremely efficient at making ATP and at storing the starting materials.

Chapter 10 on the respiratory system provides detailed information about the process of how cells convert food into ATP. This section focuses on those details of respiration that are important in order to understand how muscles get the energy to contract and what happens as muscle cells use up their energy stores.

Getting Energy from ATP

ATP belongs to a class of molecules called nucleotides—the same molecules that make up DNA. Like other nucleotides, ATP is made up of three parts: a sugar called ribose, a double ring of carbon atoms called adenine, and three phosphates attached like a tail. These phosphates are linked by high-energy bonds. When a cell needs energy, it breaks off one of the phosphates from the ATP to form adenosine diphosphate (ADP) and one free phosphate. Removing that phosphate releases enough energy to drive a cellular reaction such as flexing the myosin head. Because the remaining ADP is not a source of energy, a cell needs to constantly convert ADP back into ATP by adding a phosphate back to the tail—particularly if that cell requires a large amount of ATP to do its job, such as a muscle cell (Figure 7.6).

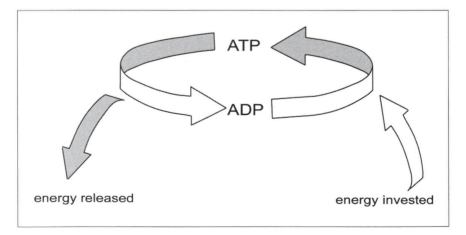

Figure 7.6 ATP and ADP. Conversion between ATP and ADP. (Sandy Windelspecht/Ricochet Productions)

It may seem surprising that breaking off a piece of a molecule can result in usable energy. However, the process is similar to how cars get energy out of burning gasoline or how fires produce heat. In each case, breaking the bonds between two atoms releases energy. In a car engine, breaking the bonds of atoms in gasoline makes energy to drive the pistons that turns the motor. In a fire, that energy takes the form of heat; and in ATP, the energy is used for cellular reactions. In each of these examples, oxygen is needed in order to break the bonds in the molecule, and carbon dioxide is released.

Creating ATP

Generating ATP is a several-step process. The first phase takes place in the main cavity of the cell, whereas the second—and most ATP-producing—step takes place in a cellular component called the mitochondria.

A mitochondrion is a small, bacteria-shaped component of the cell. Each cell has from one to many mitochondria, depending on how much energy it needs in order to survive. It is no accident that the mitochondria resemble bacteria—researchers think that hundreds of millions of years ago, free-living bacteria entered larger cells. These bacteria received food from the cell, and in return, they provided ATP. Over time, the bacteria lost their ability to exist without their host cell and became the mitochondria that cells have today. Mitochondria still have some remnants of their free-living bacterial ancestors, such as some DNA and genes that make bacterial proteins. The process mitochondria use to make ATP is similar to how free-living bacteria still function.

Glycolysis

The first step of producing ATP—called **glycolysis**—takes place outside of the mitochondria in the main compartment of the cell. This process requires a sugar called glucose as a starting material. The glucose molecule is a chain of six carbon atoms with some hydrogen and oxygen ($C_6H_{12}O_6$). It can either come from the bloodstream, where glucose constantly circulates to feed cells, or from the muscle cell's internal stores.

During the glycolysis process, a series of reactions splits glucose into two halves, each of which is called **pyruvate** (in fact, the name glycolysis means "sugar splitting"). For each sugar that enters glycolysis, the process makes two molecules of pyruvate and two molecules of ATP. Although this is a net gain in ATP, two ATP molecules are not enough to fuel extensive exercise.

Citric Acid Cycle

The next step in the ATP-producing pathway takes place within the mitochondria. Because of their critical importance in creating energy for the muscle, mitochondria are located underneath the sarcoplasmic reticulum, overlaying the muscle fibers themselves. In this position, the mitochondria are poised to receive fat or sugar from the blood and convert that into ATP. In most cells, mitochondria exist as individual units that dot the interior of the cell. They are so important to a muscle cell, however, that they fuse to form a single large entity that spans the entire muscle cell.

The process that takes place within the mitochondria is called the **citric acid cycle**—also called the Krebs cycle after Sir Hans Krebs (1900–1981), who first discovered the cycle in the 1930s. The cycle is also sometimes called the tricarboxylic acid (TCA) cycle because of the three-carbon molecule that continually moves through the cycle. The citric acid cycle begins when pyruvate from glycolysis loses one carbon atom, to become a two-carbon molecule called **acetyl coenzyme A (acetyl CoA)**. This acetyl CoA then enters the mitochondria, where it goes through eight individual steps, producing carbon dioxide that gets breathed out, one more molecule of ATP, and two high-energy molecules called NADH and $FADH_2$. These molecules have the most potential to generate ATP.

Although each acetyl CoA that enters the citric acid cycle produces one ATP, every glucose produces two molecules of acetyl CoA. This means that the citric acid cycle produces two molecules of ATP per glucose, bringing the total up to four molecules of ATP for each glucose that goes through glycolysis and then into the citric acid cycle. Again, having four molecules of ATP is better than no ATP at all, but it is not enough to sustain long-term exercise such as jogging or swimming.

Sugar is the most common fuel for making ATP, but cells can also get energy from fat in the form of free fatty acid in the blood. A fat molecule is a very long chain of carbon atoms. One common fat that is found in the cell is called palmitate, which is a chain of 16 carbons. When a fat molecule enters the cell, it is broken down in two-carbon units into acetyl CoA—the same molecule that sugar is converted into after glycolysis. This acetyl CoA then goes through the citric acid cycle and produces carbon dioxide, one ATP, and NADH and $FADH_2$.

With sugar, scientists can calculate how many ATPs come from each molecule. That is because each sugar has six carbons and sends two acetyl CoA molecules through the pathway. Fat molecules can be different lengths, so it is harder to calculate how many acetyl CoA molecules will be made from one fat molecule and therefore calculate how much ATP will be produced. Regardless, a fat molecule will produce the same amount of ATP, carbon dioxide, NADH, and $FADH_2$ per acetyl CoA as each acetyl CoA from sugar—it is just a matter of how many acetyl CoA molecules can be generated from each fat molecule.

The difference in how fat and sugar molecules enter the citric acid cycle helps explain why a gram of fat has so many more calories than a gram of sugar. A **calorie** is essentially a measure of how much energy that food contains. From each six-carbon sugar, only four carbons go through the Krebs cycle (two molecules of two-carbon acetyl CoA). The remaining two carbons are breathed out as carbon dioxide. With fat, the entire molecule is broken down in two carbon chunks, so the entire molecule is used to create ATP. Because the entire weight of the fat molecule goes to making energy, it can provide more energy per weight than sugar. This translates into more calories per gram on a food label. The benefits of burning fat over burning sugar for energy is not lost on muscle cells. As muscles become highly trained, they also become better at using fat for energy.

Electron Transport

The final phase of ATP production involves the NADH and $FADH_2$ that were made during both the citric acid cycle and glycolysis. These molecules both carry extra electrons that are in a high-energy state. They transfer their electrons to a series of proteins that are lodged in the membrane of the mitochondria. Together, these proteins are called the electron transport chain. They form a continuous path for electrons, picking up the electrons from NADH and $FADH_2$, and transferring them down the chain.

This step leaves the NADH and $FADH_2$ short on electrons. The depleted molecules go back to the citric acid cycle, where they pick up new high-energy electrons that can once again be donated to the electron transport chain.

In the final step of the electron transport chain, the electrons combine with oxygen in the mitochondria to form water. Oxygen is essential to this step. Without oxygen, electrons pile up in the electron transport chain, and transport stops. When this happens, the citric acid cycle also grinds to a halt, pyruvate stops being imported into the mitochondria, and all the ATP must be made by glycolysis.

The point of moving electrons along the mitochondrial membrane is not simply to convert an oxygen into water—at each step of the chain, the same reaction that transfers the electron to the next protein also transports hydrogen ions from inside the mitochondrial membrane to the outside of the membrane. These hydrogens build up on the outside of the membrane like water behind a dam. When the mitochondria releases the hydrogen ions back inside (through a molecule called the ATP synthetase), those hydrogens do the equivalent of the spinning wheels inside a dam. In dams, the spinning wheels make electricity that powers homes. In a cell, unleashing the dam produces about thirty-two molecules of ATP per sugar molecule. The process of producing ATP through the electron transport chain is called **oxidative phosphorylation**.

The entire process of breaking down sugar in glycolysis, sending pyruvate into the mitochondria, and transporting electrons down the electron transport chain is shown in Figure 7.5.

The Importance of Oxygen

Only 4 of 36 total molecules of ATP come from glycolysis and the citric acid cycle, while the remaining 32 molecules come from oxidative phosphorylation. Oxidative phosphorylation is a rich source of ATP, as long as the mitochondria receive a steady supply of oxygen.

The oxygen used in the electron transport chain accounts for why a person's breathing and pulse increase when they exercise. If a person cannot deliver enough oxygen to the muscles, then the electron transport chain shuts down, and the muscle receives only the paltry amount of

ATP made by glycolysis. For low-level exercise such as gardening, a person will breathe slightly harder than usual in order to provide the muscle cells with enough oxygen for their slightly increased ATP demands. A person in a swimming race, on the other hand, must breathe very hard and pump new oxygen-containing blood to the muscles very quickly to keep up with the muscle's high demand for ATP (Figure 7.7).

Exercise that takes place through ATP made by oxidative phosphorylation is called **aerobic exercise** (aerobic means "with oxygen"). Exercise that takes place when oxygen is limited and all ATP comes from glycolysis is called **anaerobic exercise** (anaerobic means "without oxygen").

Once a person is done exercising, his or her breathing and heart rate stay elevated for several minutes even though the muscles are no longer using ATP. This happens because the muscle cells need to replenish their supply of stored-up ATP. A person will continue breathing faster than

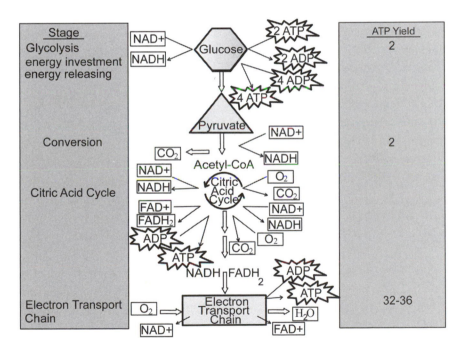

Figure 7.7 Overview of energy production from glucose. (Sandy Windelspecht/Ricochet Productions)

usual until the muscle has enough ATP stored up to act as a buffer against future bursts of exercise.

Sources of Energy

Cells in the body can convert sugars, fat, or protein from the bloodstream into ATP. Although the bloodstream provides a continuous source of fuel, muscles also store their own fuel in the form of molecules called **creatine phosphate** and glycogen (Figure 7.8).

Creatine Phosphate

Researchers knew from early experiments that a backup energy source must exist in muscles. In these experiments, Eggleton and Eggleton removed a piece of muscle but kept the nerve intact so that it could still stimulate the muscle to contract. When they gave that muscle chemicals

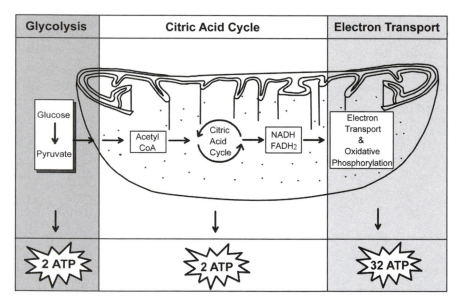

Glycolysis	Citric Acid Cycle	Electron Transport

Figure 7.8   ATP Production. ATP produced from glycolysis, the citric acid cycle, and electron transport. (Sandy Windelspecht/Ricochet Productions)

to prevent either glycolysis or oxidative phosphorylation, and then stimulated the nerve, the muscle could still contract. This result told the researchers that the muscle must have some way to regenerate ATP that did not include either glycolysis or oxidative phosphorylation.

In 1927 the researchers identified this short-term energy store as a molecule called creatine phosphate. Creatine phosphate serves as a reservoir for phosphates that can be added to ADP to regenerate ATP. When the cell is at rest and making excess ATP, then the ATP will transfer one phosphate to creatine to make creatine phosphate, converting the ATP to ADP. (The ADP receives a new phosphate through glycolysis or oxidative phosphorylation and is converted back into ATP.)

When a person begins exercising—for example, by suddenly standing up and beginning to walk briskly—the muscles do not have enough time to ramp up ATP production. These muscles will use up all the available stored ATP in about two seconds. That is less time than is needed for the muscle to begin producing energy through glycolysis and oxidative phosphorylation. During the lag time between when muscles use up stored ATP and when new ATP can be created, the creatine phosphate donates phosphate back to ADP to regenerate ATP and maintain energy for a contraction. By the time the cell depletes its store of creatine phosphate, the mitochondria have begun churning out a fresh supply of ATP.

When the exercise ends, the muscles refill the bank of creatine phosphate to be used at a later time. This refueling can happen on the order of a few minutes, which is why a sprinter or weight lifter can do several repetitions of creatine phosphate–depleting exercise with only a few minutes to recover.

Glycogen

After eating a meal, digestive enzymes in the stomach and small intestine break down food into sugars, protein, and other nutrients that are absorbed into the blood from the small intestine. For about an hour after eating a meal, the blood has a lot of sugar in the form of glucose, which the cells of the body take up and use to produce ATP or store for later use. However, the body eventually uses up all that sugar and must turn to another source of energy.

Cells of the liver and muscle save sugar to use when supplies are limited between meals. They both take up excess sugar when it is plentiful and convert it into chains of six-glucose molecules called glycogen. In the muscle, the sarcoplasmic reticulum serves as a reservoir for this glycogen. When the muscles need energy, they can convert the glycogen back into sugar and feed that sugar into the energy-producing pathway to produce ATP. The liver releases its glycogen as sugar into the blood to keep a steady supply of sugar available to other cells in the body, including the brain.

Where creatine phosphate provides a short burst of energy for cells, it is quickly depleted. After the creatine phosphate is used up, the muscles turn to glycogen stores to create new ATP. In most cases, this stored glycogen will provide enough energy for any activity the muscle needs to accomplish, such as walking, running, or housecleaning.

Fat

When blood sugar levels are high after a meal, liver and muscle cells are not the only ones planning ahead. Fat cells take up sugar and convert it into a form of fat called a triglyceride, which the cell stores as a single, large droplet for later use. This fat is a reserve against the future when the body needs fuel and all the creatine phosphate and glycogen are used up. The fat cells then release the stored fat into the blood in a form called a fatty acid, which the mitochondria can convert into ATP. In addition to fatty acids in the blood, muscles cells store tiny droplets of fat that they can use during exercise.

Although the muscles use creatine phosphate, glycogen, and fat more or less in sequence, it is important to note that the body almost always has some sugar and some fatty acid present in the blood at any time. Muscles will also use these fuel sources in different quantities under some conditions. As an example, highly trained endurance athletes rely more heavily on fat for energy—an adaptation that makes the muscle far more efficient at getting the most energy out of the lowest-weight fuel source. Moreover, the brain can only survive on sugar, so for normal brain function, there must always be some sugar in the blood.

Protein

Most food is made up of sugars (carbohydrates), fat, and protein. Although the body stores sugar and fat for later use, it does not store excess protein. Instead, the protein subunits (called amino acids) are taken up from the bloodstream and used to build new proteins in cells. Some of that protein can also be taken up by liver cells and converted into sugar that then fuels ATP production in cells or is converted into fat by fat cells.

Protein in the bloodstream is very rarely used as an energy source. During starvation, the body may break down existing proteins and use that to fuel the production of ATP. But in a normal situation, protein contributes only about 5 percent of the fuel to make ATP. Some protein subunits can be broken down into acetyl CoA and fed directly into the citric acid cycle. Other subunits feed in partway through the cycle and only produce some of the ATP, NADH, and $FADH_2$ that are normally made in the process.

Regulating the Fuel Source

Which source of energy the muscle cell uses depends on both what type of work the muscle is performing and what energy source is available. For short, intense work, the muscle will rely on creatine phosphate and muscle glycogen. But for longer work, it may rely on a combination of muscle glycogen, sugar and fatty acids from the blood, and even protein. A number of factors determine what energy source a muscle uses, including diet, caffeine, hormones, and the environment.

Sugar Availability

After exercise that uses up all of the stored glycogen—such as running a 10-kilometer race—it takes about 24 hours before the muscles can refill their glycogen supplies. If the person uses the muscles before they have replenished the glycogen supplies, the muscle will have to rely more heavily on fat for energy. People who eat a high-fat diet will have more fat than sugar available in the blood and will likely use fat more extensively to generate ATP.

Caffeine

One of caffeine's many side effects is that it causes fat cells to release fatty acids into the blood. With higher blood concentrations of fatty acids, the muscles will rely more on fat and less on stored glycogen for energy. The effect is particularly relevant in the first 20 or so minutes of exercise, when muscles usually use very little fat. By drinking caffeine before an event, athletes can increase their endurance by using fat for energy early on and sparing the muscle glycogen for later.

Hormones

A combination of hormones in the body regulates the sugar and fat supplies to ensure that the muscles have a steady supply of fuel. Hormones such as adrenaline—which the body secretes under times of stress and during exercise—cause the body to release more fatty acids and therefore to use fatty acids for energy in the muscles. Adrenaline also causes the liver to break stored glycogen into glucose and release that into the blood. Two other key hormones regulate how sugar is used by the body. The pancreas releases **insulin** after a meal when sugar levels are high. During this time, muscles take up sugar to store as glycogen and are more likely to use sugar to make energy. When sugar levels fall, the pancreas stops making insulin and instead makes glucagon. Glucagon causes the fat cells to release fatty acids and the liver cells to release glucose. During exercise, the body produces additional glucagon, which further increases the amounts of fat and sugar that are available to the muscles. For more information about how insulin and glucagon regulate blood sugar levels, see Chapter 4 on the endocrine system. (See also Sidebar 7.2.)

With so many hormones all telling the liver to release glucose during exercise, blood glucose levels can be as much as 50 percent higher after a short period of exercise. This phenomenon is in part responsible for why people often do not feel hungry immediately after exercising. For most athletic events, the liver releases glucose at about the same rate that the muscles take it up. But for very long events, the liver may run out of glycogen and not be able to keep pace with the muscle's glucose demands. At this time, the muscles rely entirely on fat or on food such as energy bars or gels.

SIDEBAR 7.2

Exercise Improves Type 2 Diabetes

To take advantage of the glucose in the bloodstream, muscles that are exercised regularly become better able to take up glucose from the blood. This effect is also why people with type 2 diabetes are encouraged to exercise regularly. In type 2 diabetes, people become increasingly resistant to insulin, which is the hormone that helps cells of the body take up sugar. With the cells not able to take up sugar, blood sugar levels increase and cause damage to the eyes, kidneys, nerves, and blood vessels. Through regular exercise, muscles in people with type 2 diabetes take up more sugar from the blood, lowering blood sugar levels and helping to prevent long-term damage to the organs. Exercise can also cause a person to lose weight, which also helps control type 2 diabetes.

Environment

Heat and altitude can both cause the body to use carbohydrates for energy and to rely more on glycolysis for ATP. At higher altitudes, the air contains less oxygen, and therefore less oxygen is present in a person's bloodstream. With limited oxygen, the muscles cannot generate enough ATP through oxidative phosphorylation and must instead use glycolysis for the remaining ATP. This effect can leave a person winded and unable to keep exercising as long as would be possible at lower altitudes.

A similar phenomenon happens in the heat. The body diverts blood to the skin where it can radiate heat, but this takes blood volume away from the muscles. With less oxygen, fatty acid, and sugar being delivered, the muscles must use stored glycogen to make energy through glycolysis. As with exercise at a high altitude, a person exercising in the heat will have less endurance than on a cooler day when the muscles generate ATP though oxidative phosphorylation.

Energy Use during Sprint and Endurance Exercise

During long-distance events such as a 10-kilometer running race, the muscles receive enough oxygen from the blood and generate ATP through oxidative phosphorylation. In one study, researchers found that marathon runners rely 99 percent on aerobic forms of generating ATP, whereas in a 100-meter sprint, 90 percent of the energy comes from anaerobic processes. An 800-meter sprint lies between these two extremes, with 60 percent of energy coming from anaerobic means.

Generating Energy in Different Muscle Types

As mentioned earlier, there are two general types of muscle fibers—slow twitch and fast twitch. The slow-twitch fibers are specialized for endurance use, while fast-twitch muscles excel at short, intense bursts of strength. With their respective specialization, the two muscle types differ in how they generate energy and how effectively they use oxygen. (See the previous section, "Energy Use during Sprint and Endurance Exercise.")

Energy in Slow-Twitch Muscles

Muscles use slow-twitch fibers for sustained exercise such as walking, dancing, or even running a marathon. These activities rarely involve large amounts of strength but do require the muscle to contract repeatedly over a long time. In order to generate enough energy to contract over such a long time period, slow-twitch muscles must be very effective at delivering oxygen so the muscles can maintain oxidative phosphorylation. As mentioned earlier, any type of endurance exercise that requires large amounts of oxygen is called aerobic exercise.

Blood vessels threading throughout the slow-twitch muscle fibers ensure that the muscle gets enough oxygen to make ATP effectively. These blood vessels bring oxygen from the lungs and return carbon dioxide created through the citric acid cycle. Blood vessels also take up water that is made in the final step of the electron transport chain. In addition to having blood vessels bring in plenty of oxygen, slow-twitch muscles contain a protein called **myoglobin** that is extremely effective at removing oxygen from the blood. This myoglobin gives the fiber a red color, which was originally used by physiologists to distinguish between fast- and slow-twitch muscle fibers.

To make the best use of all the oxygen coming in from the blood vessels, slow-twitch muscles contain many more mitochondria than fast-twitch muscles. In fact, in some muscle fibers, these mitochondria make up 20 percent of the muscle fiber volume.

Energy in Fast-Twitch Muscles

Fast-twitch muscles are used for bursts of speed—exercise that ends before the heart can begin beating faster to deliver more oxygen to the muscles. In order to generate enough ATP for the contraction without being able to rely on the electron transport chain, fast-twitch muscle fibers are adapted to be extremely efficient at glycolysis. They have fewer blood vessels and mitochondria than slow-twitch muscles and store more creatine phosphate in order to replenish the ATP supplies. Fast-twitch muscles also contain more of the molecules that carry out the process of glycolysis and store far more glycogen than slow-twitch muscles to ensure a continuous supply of sugar to feed into glycolysis.

As discussed earlier, there are two types of fast-twitch muscles—type IIa and type IIb. Type IIb fibers are the ones that are most specialized for fast, sprint-type work. They have the fewest mitochondria, least myoglobin, and most glycolysis enzymes of any muscle fiber.

The type IIa fibers are somewhere between the slow-twitch fibers and type IIb fibers. Their myosin heads pull and release the actin filaments quickly, like the IIb fibers, and they are used for sprinting rather than endurance exercise. They have more mitochondria than IIb fibers, however, and are also darker in color because they contain some myoglobin. These fibers often serve as a backup when the slow-twitch muscles become fatigued. Table 7.3 shows the differences between the different muscle types. (See also Sidebar 7.3.)

Energy Use in Cardiac Muscle

The previous material has dealt primarily with how skeletal muscles generate and use energy. However, cardiac and smooth muscles also have characteristic ways of generating ATP. Cardiac muscles resemble slow-twitch skeletal muscles in that they must continuously perform a

TABLE 7.3
Differences between Muscle Fiber Types

	Slow-twitch	Fast-twitch IIa	Fast-twitch IIb
Contraction speed	Slow	Fast	High
Number of mitochondria	High	Medium	Low
Glycolysis enzymes	Low	High	High
Citric acid cycle enzymes	High	Medium	Low
Creative phosphate levels	Low	High	High
Motor unit strength	Low	High	High
Endurance	High	Moderate	Low

low-intensity contraction. Each cardiac muscle cell has mitochondria packed between the muscle fibers where they can provide a steady supply of energy. Near the mitochondria, the heart cells also have droplets of fat that the cell uses almost exclusively for energy.

SIDEBAR 7.3

Light and Dark Meat Represent Fast- and Slow-Twitch Muscles

The differences between fast- and slow-twitch muscles are noticeable by the naked eye as light and dark meat in chicken or turkey. The dark meat on the thighs is juicier because of all the fatty membranes around the blood vessels and mitochondria. The difference in color comes from a protein called myoglobin that helps the muscle extract oxygen from the blood. In poultry, the thighs are used all day long for walking and standing, so these muscles have a large number of slow-twitch fibers and are dark in color. The lighter muscle that makes up the breast meat is used for short bursts of flying—exercise that takes a lot of strength but does not last long. This muscle is dryer because it has less fatty cell membranes surrounding the blood vessels and mitochondria and is lighter in color because it has less myoglobin.

The heart absolutely relies on having a steady blood supply to provide oxygen. During a heart attack, one or more of the blood vessels that bring blood to the heart muscles gets blocked. Even a temporary loss of oxygen can prevent the heart from making enough ATP and contracting normally, cutting off the blood supply to other parts of the body including the brain. The heart is unable to compensate for the lack of oxygen by calling on creatine phosphate stores or by making ATP through glycolysis.

Energy Use in Smooth Muscle

Smooth muscle has a different way of contracting than either skeletal or heart muscle. In these cells, the myosin heads cycle through ATP about 10 times more slowly than the myosin in heart and skeletal muscle. This makes the smooth muscle very slow to contract, but once it is contracted, the muscle burns through ATP very slowly and can sustain the contraction for a long time. This quality makes smooth muscle particularly good at squeezing, such as squeezing food through the digestive system or maintaining a blood vessel's diameter.

With its limited ATP needs, smooth muscle has few special adaptations for generating ATP. Like all the cells of the body, smooth muscle cells contain mitochondria that generate ATP from the sugar or fat circulating in the blood. Smooth muscle cells do not store their own energy supplies, nor do they have excess mitochondria. In smooth muscle cells, as in most cells of the body, the mitochondria do tend to congregate near the actin and myosin fibers where the ATP is most needed.

Muscular Adaptation to Exercise

It does not take an exercise physiologist to tell the difference between a person who works out and a person who does not. Those who get regular exercise tend to be leaner and have better-defined muscles than those who do not exercise. The lower weight is a function of burning calories while exercising. But other differences, such as stronger, better-defined muscles, are a result of molecular changes that take place within the muscle. Some of these changes are visible, but others are invisible adaptations that make the muscles and heart better able to run, bike, row, or lift weights. The muscle fiber type can change, fibers themselves grow larger, muscles contain more ATP-producing enzymes, and the muscles become better able to use oxygen. Remember that there are two types of exercise: anaerobic and aerobic. In aerobic exercise, such as biking, most of the energy comes from oxidative phosphorylation. In anaerobic exercise, such as weight lifting or sprinting, a person cannot deliver enough oxygen to the muscle and so the muscle relies primarily on glycolysis for ATP. Whether a person does aerobic or anaerobic training controls what types of changes take place in the muscle.

Muscle Adaptations to Anaerobic Exercise

Gains in Muscle Strength

Regular anaerobic exercise causes muscles to grow larger and more defined. In the past, researchers thought that the increase in muscle size led directly to an increase in strength. Keep in mind that during this time, most athletes were men who did indeed develop bulky muscles as they gained strength, and weight-lifting champions do have larger muscles than ordinary people. Together, these factors build a compelling case that muscle strength and size are related.

To some extent, these observations do hold true—the men's and women's weight-lifting champions do all have extremely large muscles. However, studies in women and children show that more is going on than just an increase in muscle size.

Even in the first eight weeks of training, when untrained men and women first begin lifting weights, the muscles become much stronger. However, during this time, their muscles do not increase noticeably in size. This is especially true of women and children, who can increase their strength the same percentage as men without showing the same increase in muscle girth. From this, it appears that muscle size does contribute to muscle strength, but that other factors also play a role in determining a muscle's strength.

Hypertrophy

When a muscle grows large in response to weight training, that gain in size is called **hypertrophy**. Immediately after exercising, a muscle is filled with fluid and feels pumped up. This hypertrophy lasts only a few hours after exercise. True muscle hypertrophy lasts as long as the muscle is in regular use, but decreases if the muscle is not used regularly.

Increases in Muscle Fiber Size

For a long time, scientists thought that people were born with a certain number of muscle fibers in each muscle and that this number could not change. If that were the case, then all muscle hypertrophy would be due to increases in fiber size, because no new fibers could be formed. One compelling reason to believe this is that scientists can look under a microscope and see larger individual muscle fibers in a person who has trained on weights for many weeks compared to that same person before they began training. The individual fibers grew larger, leading to an overall larger muscle.

The addition of new actin and myosin filaments within a muscle fiber happens because the muscles produce new protein. At any given time, a muscle is both building new protein and breaking down old protein. The balance between these processes keeps a muscle at a certain size. Regular strength training increases the amount of new protein that muscles make and decreases the amount of protein that is broken down, leading to a net gain in muscle mass.

This balance between protein being made and being broken down is also altered by the hormone **testosterone**, of which men have much more than women. Testosterone increases the amount of protein that muscles make. This accounts, in part, for why men develop larger muscles than women do. Some experiments lead scientists to think that endurance athletes rely exclusively on increases in muscle fiber size to gain their increases in strength, while sprint athletes rely very little on increases in fiber size. When researchers train animals to press a lever several times to receive food, those animals have the same number of muscle fibers before and after training, but those fibers grew larger. However, they mainly see this effect in animals who are trained to press light weights and must repeat the movement many times throughout the day—much like endurance training.

Increases in Muscle Fiber Number

Increases in muscle fiber number seem to happen mainly in athletes who train on heavy weights. When cats were trained to press a very heavy lever to get their food, researchers saw muscle fibers in the process of splitting in two, generating new fibers. They also counted the muscle fibers and found more individual fibers after training than before, probably as a result of fibers splitting to create additional fibers.

Although it is hard to count the exact number of fibers in a person's muscle, researchers can look at the overall size of muscle fibers. When they compare muscle fibers in bodybuilders and people who are fit but have done no weight training, they find that the muscle fibers are about the same size. Because the bodybuilder's overall muscles are much larger, the researchers conclude that the bodybuilders must have more fibers in total.

One reason that strength training increases the number of muscle fibers may have to do with how the muscles respond to lifting heavy weights. After a heavy weight-lifting session, muscle fibers have some damage due to the stress of lifting heavy weights. In the process of repairing those damaged cells, new fibers can be formed.

Strength Increases Due to Exercise

Scientists are not precisely sure how muscles become stronger without becoming bigger, such as in the case of women or children or during the early weeks of training in men. Several studies have led exercise physiologists to believe that changes in the nervous system may explain these increases in strength. In one study, participants did strength-training exercises with one arm for eight weeks. At the end of the study, the participants had grown 25 percent stronger in the trained arm, as expected. But these people were also about 15 percent stronger in the untrained arm. This result tells researchers that factors in addition to muscle strength must have changed in response to exercise.

One explanation is that when the muscle of a trained person contracts, more motor units contribute to the contraction. Because a person is using the entire muscle, he or she can lift a heavier weight. This could also explain how people pull off remarkable feats of strength when they are under pressure, such as lifting people out from under cars or freeing themselves when pinned under heavy objects. The muscle may contract all motor units at one time, making the muscle significantly stronger than when only a few motor units contribute to a contraction.

Another explanation could be that when a trained muscle flexes, the opposite muscle does not resist as much. For example, when a trained person does an arm curl using the biceps muscle, the opposing triceps muscle relaxes and allows the biceps to flex (see the earlier discussion in this chapter of how muscles work in antagonistic pairs). According to this theory, the triceps muscle in an untrained person does not relax as much, so it takes more strength on the part of the biceps to overcome the resistance. If this explanation were true, then training with one arm causes the brain to alter how it instructs the triceps of both arms.

Scientists have also seen changes in the junction between the nerves and the muscle in people who are highly trained. Although they still do not know how those changes relate to muscle strength, it is possible that these changes allow the muscle to contract more strongly in response to a given nerve signal.

At this time, scientists do not have enough evidence to figure out which explanation is right. It could be that all of these changes take place to some extent, each contributing to the overall gain in strength.

Muscle Adaptations to Aerobic Exercise

Increased Ability to Use Oxygen

Whereas anaerobic exercise such as weight lifting or sprinting causes the muscles to grow larger, aerobic exercise causes changes in the way the muscle uses energy. Keep in mind that aerobic exercise such as swimming or rowing does not use a muscle's full strength, but that muscle must be able to contract repeatedly over a long time. The muscles will become stronger and larger than in a person who does not work out at all, but they never achieve the size or strength of a weight lifter's muscles. The hurdle for endurance athletes is delivering enough oxygen to the muscles so the muscles can make ATP for contraction. Sidebar 7.4 takes a closer look at when exercise-related muscular injuries.

SIDEBAR 7.4

When Exercising Hurts

With obesity-related diseases on the increase in the United States, more people are staying active and exercising regularly. But according to the National Institute of Arthritis and Musculoskeletal and Skin Diseases, more people are now suffering sports injuries. Those people who overdo a certain exercise or who do not train or warm up properly are particularly vulnerable to sports injuries.

Most of these injuries can be treated effectively, and physical activity can resume with proper rest and recovery. In fact, many sports injuries can be avoided with the proper prevention measures.

Common sports injuries related to the muscular system often involve sprains and strains, as well as compartment syndrome and Achilles' tendon injuries. Here are some details on each of these injuries:

Sprain: This injury occurs when there is a stretch or tear in a ligament— a band of connective tissues that connect the ends of bones. Sprains can range in severity. Ligaments can be simply stretched, resulting in inflammation, tenderness, and/or pain. In severe cases, ligaments can be torn, which can make it difficult to move the affected limb or joint. The most areas of the body most vulnerable to sprains are the ankles, knees, and wrists.

Strain: This injury involves a muscle or tendon—the connective tissue joining muscles to bones. Strains occur when the muscle or tendon is pulled, twisted, or torn as a result of overstretching or overcontraction. Symptoms of this injury include pain, muscle spasm, and loss of strength.

Compartment Syndrome: A compartment is composed of tough membranes located throughout the body. Within these compartments are muscles, nerves, and blood vessels. When a muscle is injured and becomes inflamed, the swollen muscles can expand, filling the compartment. This swelling can cause interference with the nerves and blood vessels in the compartment, resulting in a painful condition known as compartment syndrome. This can be caused by an acute hit to certain parts of the body, or ongoing overuse injuries that can occur in sports such as long-distance running.

Achilles Tendon Injuries: The Achilles tendon connects the calf muscle to the back of the heel. Injuries in this area can occur when the tendon is stretched too far, torn, or irritated in some manner. Achilles tendon injuries can cause severe and sudden pain. These injuries often occur among people who do not warm up properly prior to exercising. There have also been studies linking Achilles tendon injuries to professional athletes who are required to quickly accelerate and jump, such as in football and basketball.

Blood Supply

Capillaries run throughout the muscle fibers, delivering oxygen and nutrients to the muscles. As people train for endurance events, their muscles accumulate more capillaries to meet the energy needs. Over time, trained people can have as many as 15 percent more capillaries in their muscles than untrained people.

Myoglobin

Remember that myoglobin in the muscle takes oxygen from the blood and delivers it to the mitochondria. An increase in capillaries will not do the muscle any good unless that muscle contains enough myoglobin to carry the increase in oxygen. The slow-twitch fibers that are most commonly used for endurance activities already contain more myoglobin than fast-twitch muscles. With training, slow-twitch fibers will contain as much as 75 percent more myoglobin than untrained fibers.

Mitochondria

The mitochondria produce ATP for the cell, using oxygen to manufacture ATP and water. The amount of ATP that can be made in a muscle cell depends, in part, on how many mitochondria that cell contains. As people do aerobic training, both the number of mitochondria within their slow-twitch muscle fibers and the size of those mitochondria increase.

Fast-twitch type IIa fibers normally contain some mitochondria, though not as many as in slow-twitch fibers. With aerobic exercise, these type IIa muscle fibers also accumulate more, larger mitochondria. This increase helps those fast-twitch type IIa fibers contribute to the aerobic exercise.

Enzymes

Within the muscle cells, sugars go through glycolysis, which produces pyruvate. This pyruvate is converted into acetyl CoA, which then enters the mitochondria where the citric acid cycle uses that acetyl CoA to generate NADH and $FADH_2$, which eventually leads to ATP production. The end result—producing ATP—relies on the citric acid cycle working at optimal speed. To make sure the cycle can produce enough NADH and $FADH_2$ to meet a muscle fiber's energy needs, muscles in trained athletes contain more of the enzymes that are used during the citric acid cycle. The activity of these enzymes increases throughout aerobic training. People who get even a small amount of aerobic exercise have much more enzyme activity in their muscles than in people who do not train. Highly trained people can have as much as double the citric acid cycle enzyme activity of untrained people.

Changes to the Heart Muscle during Exercise

With all the changes that take place to skeletal muscle as a result of training, it is easy to overlook the heart muscle. However, the heart plays a key role in a person's exercise performance and also undergoes some changes through exercise. These changes help the heart pump more blood to the muscles. The regular blood supply is critical—the blood brings oxygen and fuel to the muscle and clears waste and lactic acid. Without sufficient blood flow, the muscles would not be able to maintain their peak contraction.

Heart Size

Through regular training, the heart muscle becomes larger, as do skeletal muscles. The heart is divided into four chambers, of which one (called the left ventricle) is primarily responsible for pumping blood containing oxygen out to the body. The muscles surrounding the left ventricle grow larger in endurance-trained athletes, and the chamber cavity grows larger to accommodate more blood. Resistance-trained athletes also have some increase in left-ventricle muscle thickness, though the chamber size remains about the same as in untrained people. The thickness of a person's heart muscle directly correlates to their **VO_2max**—endurance-trained athletes with a high VO_2max also have thicker heart muscle walls, whereas sedentary people with a low VO_2max have thinner heart muscles.

A thicker heart muscle means a stronger contraction, pushing blood out to the muscles where it is needed. Many athletes notice this change in heart muscle strength by a lower resting heart rate. Because the heart muscle is stronger, it pumps more blood with each contraction. At rest, the heart can beat fewer times per minute and still distribute the same amount of blood as a weaker heart pumping at a fast rate. Untrained people often have a resting heart rate of between 65 to 80 beats per minute, though extremely sedentary people can have heart rates as high as 100 beats per minute. Trained athletes have been known to have resting heart rates as low as 28 to 40 beats per minute.

The larger chamber size also contributes to an athlete's lower heart rate. Trained athletes have more blood volume than sedentary people, so the chamber fills fuller per beat than in untrained people. The chamber also has more time to fill between the slower beats, further increasing the blood volume in the heart chamber. With these factors combined, athletes force much more blood out to the body with each beat than untrained people. In one study, the total volume of blood forced out by each heartbeat went up by almost 50 percent after a six-month training regime.

Heart Rate during Exercise

During exercise, the heart rate increases to send more blood to the muscles. When an athlete reaches a steady pace, such as the pace for a two-hour bike ride, the heart rate levels off and will stay about the same for the given amount of exercise. This heart rate is called the steady-state heart rate. If the athlete increases or decreases the pace, the steady-state heart rate will also increase or decrease.

People who are in good physical condition will generally have a lower steady-state heart rate for the same amount of exercise as a person who is untrained. This lower heart rate is a result of the heart squeezing out more blood volume per heartbeat in trained people.

For all-out exertion, that steady-state heart rate will reach a maximum that the heart cannot exceed. A person can reach the highest maximum heart rate at about age 10 to 15, and that maximum decreases by about one beat per minute per year. A person's maximum heart rate can be approximated with the formula 220 − age (in years). So, a 40-year-old person would have a maximum heart rate of 220 − 40 = 180, or 180 beats per minute. This formula is simply an estimate—many people have heart rates that fall outside the calculated maximum.

The decrease in maximum heart rate seems to happen as a result of how the signal to contract spreads across the heart. Even with regular exercise, which causes the muscle to contract with more force, a person's maximum heart rate will decrease with age. This does not generally affect a person's athletic performance for endurance-type events, because in these events, the heart rate rarely reaches its maximum. However, as people age, they may notice their sprint performance declining. People reach their maximum heart rate when sprinting—if the maximum heart rate is lower, a person will tire faster when sprinting.

Changes to Smooth Muscle during Exercise

When a person is resting in a chair, the muscles receive about 20 percent of the total blood flow, with the liver, kidney, and brain making up about 60 percent of the blood flow and the remaining blood distributed through the heart muscle, skin, and other tissues. During heavy exercise, the heart muscle continues to receive about 5 percent of the blood flow, while the skeletal muscle receives 80 to 85 percent of the total blood flow and the other organs receive severely restricted blood flow. This massive redistribution of blood arises due to changes in the smooth muscle lining that surrounds blood vessels.

Blood leaves the heart through large vessels called arteries. These arteries branch out, sending offshoots to all regions of the body. When the blood reaches a tissue, such as a muscle or the kidneys, the arteries branch into a network of tiny vessels called capillaries. These capillaries then feed into veins that return blood from the organs back to the heart. All of these vessels are surrounded by sheets of smooth muscle that regulate the size of the vessel. When the smooth muscles contract, the vessels become narrow and carry less blood.

Several signals direct when muscles surrounding blood vessels should contract. One of these is the level of oxygen in the surrounding tissue. Remember that muscles use up more oxygen as they increase their demand for ATP. Where oxygen levels are low, blood vessels open up to allow more blood into the region to supply fresh oxygen and fuel.

At the beginning of exercise, signals throughout the body cause smooth muscle surrounding capillaries in the digestive system and other organs to constrict, reducing blood flow. At the same time, signals such as lower oxygen levels instruct capillaries in the muscle to expand, allowing more blood to the region. As exercise continues, more and more blood is redirected to the muscles. The redistribution of blood to the muscles is dependent, in part, on the blood supply needs of other tissues. For example, exercise too soon after a meal can result in not enough blood being directed to the muscles. In studies in both humans and other animals, eating a meal directly before exercise caused a 15–20 percent drop in blood flow to the muscles. This blood was redirected to the intestinal tract, where it was needed to help digest food. Many athletes avoid eating soon before exercise in order to have as much blood flow as possible available to the muscles.

Summary

There are three types of muscles in the human body. The first, the skeletal muscle, helps the body move, as they are attached to the skeleton; the second, smooth muscle, is an integral part of the internal muscles. The third type of muscle, cardiac, works to help the heart to function. Each of these muscles contract differently, and based on messages from different areas of the body. The smooth and cardiac muscles are directed by the autonomic nervous system, while the skeletal muscles respond to the central nervous system.

Muscles require energy to function, and interacts with nerves to contract. Specifically, two types of muscular nerve fibers determine how muscles perform—fast-twitch and slow-twitch fibers. Muscles produce energy and take energy from various sources, including foods and drinks. Muscles also help us to maintain a healthy body through regular exercise. However, muscles adapt differently to anaerobic and aerobic exercising.

8

The Nervous System

Julie McDowell

Interesting Facts

- The average weight of an adult human brain is between 2.8 and 3.1 pounds (1,300 and 1,400 grams).

- The average weight of an elephant's brain is 17.2 pounds (7,800 grams).

- The average number of neurons in the brain is 100 billion.

- The length of myelinated nerve fibers in the brain is between 93,200 and 112,000 miles (150,000 and 180,000 kilometers).

- The difference in the number of neurons in the brain's left and right hemispheres: 186 million more neurons in the left hemisphere in comparison to the right hemisphere.

- Total surface area of the human brain's cerebral cortex: 2.5 square feet (2,500 square centimeters).

- Total surface area of an elephant's cerebral cortex: 6.8 square feet (6,300 square centimeters).

- Total number of neurons in the cerebral cortex: 10 billion.

- Total volume of cerebrospinal fluid in the human body: 4.2–5.1 fluid ounces (125–150 milliliters).

- Half life of cerebrospinal fluid: 3 hours.

- Nerve impulses can travel from the brain at speeds up to 170 miles (274 kilometers) per hour.

Chapter Highlights

- Central nervous system

- Nerve cells

- Cells and energy production

- Synapses

- Neurotransmitters

- Nerve impulses

- Spinal cord and spinal nerves

- The brain: lobes and the cerebral cortex

- Peripheral and autonomic nervous system

- Sympathetic nervous system

- The senses

Words to Watch For

Acetycholine

Adaptation

Adenosine diphosphate (ADP)

Adenosine triphosphate (ATP)

Adrenaline

Aerobic cell respiration

Afferent nerves

Autonomic nervous system

Brain

Brain stem

Cell body

Central nervous system

Cerebellum

Cerebral aqueduct

Cerebral cortex

Cerebrospinal fluid

Cerebrum

Choroid plexus

Chromosomes

Circadian rhythm

Convergence

Corpus callosum

Cranial nerves

Craniosacral division

Cranium

Cutaneous senses

Cytoplasm

Depolarization

Divergence

DNA

Dopamine

Dorsal root

Dorsal root ganglion

Effector

Efferent nerve

Efferent neuron

Endocrine system

Equilibrium

Excitatory nerve

Excitatory synapse

Extracellular fluid

Glucose

Gray matter

Growth hormone-
   releasing hormone

Gyri

Hypothalamus

Hypoxia

Inhibitory nerve

Inhibitory synapse

Intermediolateral cell
   column

Intracellular fluid

Medulla

Meninges

Mitochondria

Muscle spindle

Myelin sheath

Nerve

Nerve fiber

Nerve plexus

Nerve tracts

Neuroglia

Neurolemma

Neurons

Neutrotransmitter

Node of Ranvier

Noradrenalin

Norepinephrine

Nucleus

Occipital lobes

Oligodendrocytes

Organelles

Parasympathetic
   division

Peptides

Peripheral nervous
   system

Polarization

Pons

Postganglionic
   neuron

Postsynaptic process

Preganglionic neuron

Presynaptic process

Projection

Reflex

Repolarization

Sarcolemma

Schwann cells

Sensory nerve

Sensory neuron

Serotonin

Somatic neuron

Spinal nerves

Spinal reflex

Stimulus

Stretch flex

Sulci

Sympathetic division

Synapatic gap

Synapse

Thalamus

Threshold level

Ventral root

Visceral neuron

Visceral organs

White matter

Introduction

The human body's nervous system is divided into two parts: the **central nervous system** (CNS) and the **peripheral nervous system** (PNS). The CNS consists of the brain and the spinal cord, and the PNS consists of the **cranial nerves** (the brain's 12 pairs of nerves) and the **spinal nerves** (31 pairs of nerves associated with the spinal cord). Also included in the PNS is the **autonomic nervous system** (ANS), which controls the "automatic" or involuntary movements of the body's smooth muscles (found in the walls of tubes and hollow organs), cardiac muscles, and glands (see Figures 8.1 and 8.2 for a general overview of the organization of the nervous system and the somatic nervous system that controls voluntary movement). The two divisions of the ANS are the **parasympathetic division**, which dominates and controls the body during nonstressful situations, and the **sympathetic division**, which dominates and controls the body during stressful situations.

Information is sent from the PNS to the brain, which serves as the activity headquarters of the CNS. Through the five senses (sight, smell, touch, hearing, and taste), the CNS detects a **stimulus**, which is a change that prompts a response in a living organism. The brain then processes the transmitted information and initiates the appropriate response or responses in the **effector**. An effector is an organ, such as a muscle or gland,

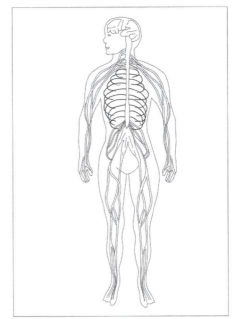

Figure 8.1 The Nervous System. The brain and nerve systems, including the spinal cord, make up the nervous system. While regulating movements, the nervous system also works to interpret sensory information. (Sandy Windelspecht/ Ricochet Productions)

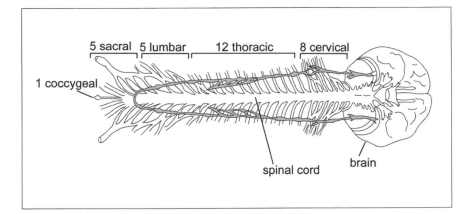

Figure 8.2   The Spinal Cord. There are five categories of spinal nerves noted on top of the figure: the coccygeal, sacral, lumbar, thoracic, and cervical, which total 31 pairs of nerves. (Sandy Windelspecht/ Ricochet Productions)

that responds through some kind of reaction (usually movement) when receiving a stimulus.

A **nerve** is defined as a group of nerve fibers located outside the CNS, and bundles of nerve fibers located within the CNS are called **nerve tracts**. Nerve fibers make up nerve cells, also known as **neurons**, which are the building blocks of the entire nervous system. Neurons are the essential conducting unit of the entire nervous system.

Before looking at the details of the neuron, it's important to understand some other important elements that enable the CNS to operate smoothly. Taking a close look at a slice of the spinal cord, one sees that most of it is composed of white material and some of it is composed of a gray substance. These two parts are simply called **white matter** and **gray matter**. Whereas the gray matter consists mostly of nerve cell bodies, the white matter is composed mostly of nerve fibers.

Nerve Cells: Foundation of the Nervous System

During the initial stages of growth, nerves develop in the embryo's CNS and then grow out and spread through the body like tentacles. There are thousands of nerve fibers grouped in large bundles that run to and from the CNS. **Afferent**

nerves are those fibers coming to the CNS from the muscles, joints, skin, or internal organs, whereas those leaving the CNS to travel to these areas are known as **efferent nerves**. The spinal cord has two types of afferent nerves: those coming in at the back are called posterior or **sensory**; those leaving the spinal cord come out the front and are called anterior or motor. When these afferent nerves reach the spinal cord's white matter, they divide and branch out to bring their messages to various different areas of the cord (Figure 8.3).

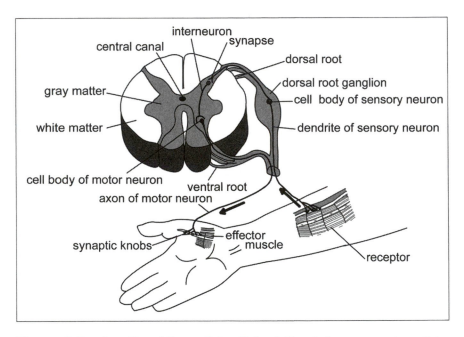

Figure 8.3   Another View of the Spinal Cord. A cross section of the spinal cord displays the white matter, which is the nerve tissue that contains myelinated axons and interneurons, and the gray matter, which is the nerve tissue that contains the neurons' cell bodies. The spinal roots—the dorsal and ventral roots—are just shown on the right side of the spinal section, although these roots and the corresponding nerve tracts are actually present on both sides of the spinal cord. This figure also shows the path of a reflex arch, with the receptor muscles receiving the impulse, which is processed by the sensory neuron and then transmitted to the motor neuron at the synapse in the spinal cord. (Sandy Windelspecht/Ricochet Productions)

Most of the branches connect with neurons near the entry region of the nerve fiber and intermediate neurons. These intermediate neurons then connect with the motor neurons, which control muscle movement.

Although neurons vary in size, shape, and functions, they all consist of the following four distinct parts:

1. *Cell body.* This is the main mass of the cell and contains the **nucleus** and other **organelles**. The nucleus is the cell's largest organelle, which contains **chromosomes**. In these chromosomes, genes carry the body's hereditary information in the **DNA**, which allows the cell to reproduce. Neuron cell bodies are found in the CNS or close to it in the trunk of the body so they are protected by bone. The **cell body** also contains **cytoplasm**, and an abundance of **mitochondria**, which are organelles responsible for energy production in the cell.

2. *Dendrites.* This group of branching nerve fibers carries impulses to the cell body. Small black spots also appear on the **dendrites**: these represent the nerve endings of other neurons, which pass messages from other neurons.

3. *Axon.* This single nerve fiber carries impulses away from the cell body and the dendrites. In humans, some **axons** can be around one meter long, but in larger mammals such as whales and giraffes, these would be much longer. A **nerve fiber** is the neuron including the axon and the surrounding cells. These fibers branch out at the neuron's ending (also known as arborization) and can be classified as either **excitatory** or **inhibitory**.

4. *Transmitting region.* The axon carries the impulses to the transmitting region, and from there it leaves the cell body and travels to the CNS.

In addition to these four parts, neurons are composed of layers of membranes and microtubules that produce hormones, neurotransmitters, and substances such as **peptides** and proteins, which will be discussed in detail further in this chapter.

How Cells Produce Energy

Cells produce energy through **aerobic cell respiration**. The following equation provides a simple explanation:

$$\text{Glucose } (C_6H_{12}O_6) + 6O_2 \rightarrow 6CO_2 + 6H_2O + ATP + \text{heat}$$

Each product of cell respiration is vital to the body's function. The heat regulates body temperature, the water feeds the cells, and the CO_2 is the waste that is exhaled through breathing.

An important product of cell respiration is **ATP (adenosine triphosphate)**, a **neurotransmitter** that is the muscle's direct source of energy for movement. Basically, ATP captures energy from food, breaks it down, and then releases it into cells. Some of this energy resulting from respiration is used by the cell's mitochondria to produce ATP. Therefore, cellular respiration is a constant, continuing cycle. All cells contain molecules of **ADP (adenosine diphosphate)** and phosphate. When the body digests food, it breaks down various chemical components to use in cell processes. A form of sugar known as **glucose** $(C_6H_{12}O_6)$ is one such substance, and is a necessary component (along with oxygen) in aerobic respiration. Glucose breaks down into CO_2 and H_2O (water), along with a release of energy. This energy then bonds with the ADP and a third phosphate to form ATP. Energy for cell processes is released and available for use when the bond holding this third phosphate is once again broken down.

Neurons

As mentioned earlier, neurons are nerve cells that are composed of nerve fibers. Neurons are the foundation of the entire nervous system, and are responsible for transmitting messages throughout the body. In the PNS, the neurons are constantly carrying information to and from the CNS. However, neurons can only carry electrical impulses in one direction, making it impossible for impulses to run into each other and cancel each other out.

The nervous system is made up of millions of neurons, in addition to the other cells that help support the functions of the spinal cord and brain. Neurons are classified into three groups: sensory, motor, and interneurons. Sensory and motor neurons make up the PNS, and interneurons are found in the CNS.

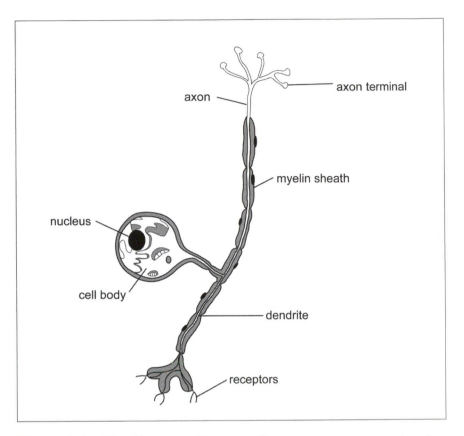

Figure 8.4   The Sensory Neuron. This neuron structure details a sensory neuron, which carries impulses and messages to the spinal cord and brain. These sensory neurons, which have cell bodies, are located in the CNS and close to the body so they are protected from damage. The axon moves these messages away from the nucleus and cell body, and the dendrites work in reverse to bring the impulses into the cell body. The myelin sheath works as an insulator to protect the neurons from short-circuiting with each other. (Sandy Windelspecht/Ricochet Productions)

The **sensory** or **afferent neuron** mainly functions in the CNS, and the **motor** or **efferent neuron** mainly functions in the PNS. Although both these neurons contain the same physical parts (the cell body, dendrites, axon, and transmitting region), there are some important differences in their composition (Figures 8.4 and 8.5).

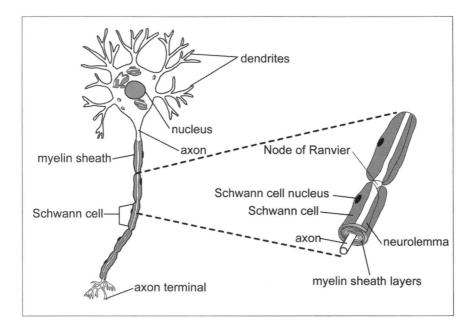

Figure 8.5   The Motor Neuron. This neuron structure details a motor neuron, which carries the impulses from the central nervous system to muscles and glands. Compared to the sensory neuron, this structure features the cell body at the top (rather than the side) and does contain receptors. (Sandy Windelspecht/Ricochet Productions)

The sensory or afferent neuron carries impulses and messages to the spinal cord and brain. In these neurons, the dendrites act as receptors—they detect external or internal changes, and transmit this information to the CNS. The sensory neuron dendrites may be short or long (sometimes as long as one meter), but they are single and are not branchlike in appearance like in the motor neuron.

Once the information is transmitted to the brain and spinal cord, the impulse is interpreted and the CNS stimulates a sensation. Sensory neurons located in the skin, skeletal muscle, and joints are known as **somatic**, and those in the internal organs are called **visceral**.

The motor (or efferent) neuron carries the impulses from the CNS to muscles and glands, also known as effectors. After the CNS processes an impulse, the motor neuron will tell muscles to contract or relax and tell glands to secrete. Those motor neurons associated with the skeletal muscle are called somatic, and those associated with the smooth muscle, cardiac muscle, and glands are called visceral.

Some axons in both in both the CNS and PNS are layered with a covering called the **myelin sheath**. Composed of fatty material, the myelin sheath electronically insulates neurons from one another. Without the protection of the myelin sheath, the neurons would short-circuit and thus be unable to transmit electronic impulses.

In the motor neurons, the axons and dendrites are wrapped in specialized cells called **Schwann cells**, which create their myelin sheath. The nucleus and cytoplasm in the Schwann cells are collectively called the **neurolemma** and are located outside the myelin sheath, physically covering the nerve cell. Schwann cells are located in the PNS. The specialized cells in the CNS are called **neuroglia**. The **node of Ranvier** is the space that separates the Schwann cells. These nodes are responsible for depolarizing electrical impulses, which will be explained in greater detail later in this chapter.

The neurolemma is important for nerve regeneration. If a nerve or nerves in the PNS is damaged or severed (for example, if a limb is damaged or severed) and then reattached through surgery, the neurolemma allows the axons and dendrites to regenerate and reattach to their proper connections. In addition, the Schwann cells are believed to produce a chemical substance that stimulates regeneration. This regeneration may be a slow process (it could take months or years), but eventually the nerve fibers may restore their functions, therefore reinstating feeling and movement to the limb.

Regeneration, however, is not possible in sensory neurons of the spinal cord because there are no Schwann cells. In these neurons, the myelin sheath is formed by the **oligodendrocytes**, the specialized cells found only in the brain and spinal cord. Because there are no Schwann cells, there is no neurolemma, and thus cell regeneration is impossible. This is why spinal cord damage or severing results in permanent loss of function.

The final classification of neurons is called interneurons. These are located entirely within the CNS and combine or integrate the sensory and motor impulses. Some of their functions include thinking, memory, and learning. For instance, interneurons might receive impulses from the brain and transmit them to the somatic nervous system that determines movement in voluntary muscles, such as fingers and toes.

Cranial nerves, or the nerves located in the brain, contain sensory, motor, and mixed nerve fibers. The sensory nerves carry impulses toward the brain, and the motor nerves carry impulses away from the brain. But most of the cranial nerves and all of the spinal nerves are made up of both sensory and motor fibers, and these are called mixed nerves.

Synapses

Every neuron has dendrites that connect with other neurons. In the CNS, neurons communicate with each other when the axon of one neuron comes into contact with the cell body or dendrite of another neuron. The space or junction between the axon and dendrite of these two neurons is known as a **synapse**, which comes from the Greek word meaning "to clasp." This is where the message carried by the neuron is passed on. The synapse is often called a relay because it is here where the information is relayed to the next neuron. More specifically, the actual area (which is between 10 and 50 nm in width) between the axon and dendrite is known as the **synaptic gap** or cleft.

Like a blueprint showing the floor plan and layout of a house, the location and pattern of connections between neurons determine the structure and organization of the CNS. The position of the synapses dictates the route that impulses will follow within and between the brain and spinal cord. The impulse pathways determine what sensations are experienced and how an effector responds to these sensations.

Events related to the synaptic process are classified as either **presynaptic** (before transmission) or **postsynaptic** (following transmission). It is important to note that neurons conduct impulses in only one direction, depending on if they are afferent or efferent.

The synapse transmissions in the CNS are extremely complicated, but Figure 8.6 describes their role in transporting information throughout the CNS at the most basic level. In this illustration are four neurons—A, B, C, and D. A and B are the presynaptic neurons that are carrying information to C and D, the postsynaptic neurons. The A and B axons will make contact with the C and D dendrites, thus creating four synapses.

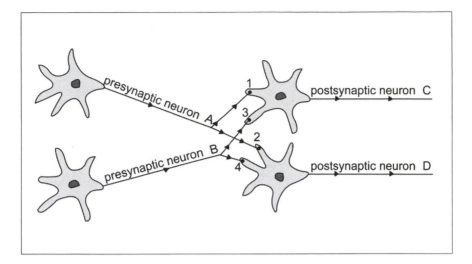

Figure 8.6 Synaptic Connections. The synapse transmission between the neurons of the CNS. The two presynaptic neurons, A and B, carry impulses and messages to the postsynaptic neurons, C and D. The locations where A and B make contact with C and D are known as synapses, which are numbered 1–4. (Sandy Windelspecht/Ricochet Productions)

Neurons connect with each other in two ways: (1) a neuron receives impulses from a few other neurons and relays these impulses on to thousands of other neurons—**divergence**; and (2) a neuron receive impulses from the nerve endings of thousands of other neurons and transmits its message to only a few other neurons—**convergence**. Convergence and divergence are two types of pathways that impulses can take when traveling from the presynaptic neuron to the postsynaptic neuron, as shown in Figure 8.6. An impulse can travel across synapse 1 to C, or it can go across synapse 2 to D.

However, the impulse can also travel along A and cross synapses 1 and 2 to C and D, which is divergence. Convergence is when the impulse travels along to presynaptic neurons to one postsynaptic neuron; for example, if A and B impulses would travel through synapse 1 and converge to postsynaptic neuron C. Of course, in the complex body, any postsynaptic neuron can expect to have hundreds of axons from presynaptic neurons attach to its dendrites.

A nerve impulse enters the cell body by first attaching itself to that cell's dendrites. This then alters the neuron's excitability. The excitatory nerve fiber amplifies the energy from the impulse, but the inhibitory nerve fiber reduces this energy. Although the arrival of the impulse alters the excitability of the neuron, it does not necessarily result in a postsynaptic neuron reaction or firing. A reaction occurs only when the nerve fiber's axon has reached its **threshold level**. The value of this threshold varies with each nerve fiber and depends on the composition of the cellular fluid and the number of impulses recently received and conducted.

For example, when one touches a warm stove with his or her fingers, the nerve impulse translating the corresponding sensation of warmth attaches itself to the dendrites of the neurons located in the fingertips, thus stimulating the neuron's excitability. The excitatory and inhibitory nerve fibers process this sensation through amplifying but also inhibiting the energy from the impulse. When the sensation (or warmth) increases beyond the threshold level of the nerve fiber's axon, a reaction occurs. So if the heat on the stove increases beyond a certain level, the nerve fibers will stimulate a chain of events that produce a reaction, such as pulling the fingers away from the stove.

Both excitatory and inhibitory neurons are managed by excitatory and inhibitory nerve fibers. Some postsynaptic neurons fire off most of the time, whereas others fire off less frequently. For example, neurons that work the muscles in the brain and spinal cord are in a constant state of excitability. Because the body's muscles must always be ready for action, there must be a constant flow of impulses between the brain and spinal cord. But neurons that control breathing are not constantly firing—they operate at a more rhythmic pace.

The respiratory and muscular responses are examples of postsynaptic excitation or inhibition because they take place immediately after an impulse transmission at the synaptic gap. Presynaptic excitation or inhibition occurs before the impulse reaches the synaptic gap. If an impulse is transmitted along an inhibitory nerve fiber, it will not deliver its full load of excitability at the synapse. In fact, this nerve fiber can completely block all nerve impulses from reaching the synapse.

Synapses can deliver messages either to thousands of neurons or to only a few. The more synaptic exchanges there are in a transmission, the more opportunities there are for changes or modifications that can be made to that resulting reaction. For example, suppose there are two ways to get from Anytown, USA, to Nowhere, USA. The first route is a straight road with no turnoffs, but the second route has numerous crossings and connects with secondary roads that ultimately lead to Nowhere. Although the first route is faster, there is no chance for changes or modifications. One cannot change the route—it is a straight shot. But the second route, even though it might be slower, has numerous opportunities for change with all its turnoffs and back roads.

When a nerve impulse crosses the synaptic gap, each synapse contains substances to create adhesion between the presynaptic and postsynaptic membranes. When contacted by the impulse, the synaptic gap constricts as the width decreases and the concentration of the transmitter increases. Synapses that are not used often tend to cease functioning, but synapses that are used frequently tend to transmit impulses quickly.

Neurotransmitters

In the latter part of the nineteenth century, there was a controversy over whether the vertebrates had a nervous system composed of a continuous network of neurons or of separate neurons. When it was eventually proven that the vertebrate's body is made up of separate neurons connected with synapses, the question remained about how the neurons related to each other.

Because nerve fibers are minute and cover only a small portion of the postsynaptic neuron's cell body, the amount of impulse and energy it can deliver is diminutive. This might be adequate for organisms such as fish or crustaceans—enough current is delivered to inhibit or excite a postsynaptic neuron in these animals. However, in reptiles and mammals, another mechanism is used to transmit impulses. In the early 1900s, scientists and physiologists had concluded that the endings of nerve fibers emit chemical substances that influence the behavior of postsynaptic neurons.

Researchers noticed that by injecting certain nerve tissue with a substance called **adrenaline**, the sympathetic system (the motor nerve network that operates the blood vessels and sweat glands, along with some of the internal organs) was stimulated. They believed that the body was constantly tapped into a natural supply of adrenaline.

These landmark discoveries verified that chemical substances are emitted through nerve endings to help transmit messages. These chemicals are called neurotransmitters, or simply transmitters. In the human body, there are about 80 different neurotransmitters.

Neurotransmitters are classified into four groups:

- *Amines*: acetylcholine, noradrenalin, serotonin, dopamine

- *Amino acids*: glutamic acids, gamma aminobutyric acid (GABA)

- *Purines*: adenosine triphosphate (ATP)

- *Peptides*: enkephalins, dynorphin, endorphin

Some important neurotransmitters in our body include **acetylcholine**, **dopamine**, **norepinephrine**, and **serotonin**. Many neurotransmitters have a particularly wide distribution and vital roles. **Noradrenalin**, for example, transmits neurons from various regions in the brain, such as the cerebral hemispheres, the cerebellum, and the spinal cord. Noradrenalin increases the reaction excitability in the CNS and the sympathetic neurons in the spinal cord.

Another major transmitter is serotonin, which is an important distributor for the sensory channels in the CNS and in the expressions of emotion. Medications that alter mood or behavior, such as antidepressants, will be targeted at serotonin and norepinephrine and will affect synapse transmission.

Dopamine is an important transmitter in the motor system, limbic system, and the hypothalamus. Parkinson's disease kills the cells that produce these transmitters, therefore impeding mobility and other motor functions. These transmitters are supplemented using a drug called L-dopa, which improves posture and motility, although it cannot stop the tremors that accompany Parkinson's disease.

In simplistic terms, when a presynaptic neuron receives an electrical impulse, its axon releases a neurotransmitter, which then diffuses across the synapse and combines with the dendrites of the postsynaptic neuron. This generates an electrical impulse, which is then carried to the postsynaptic neuron's axon to the next synapse and so on. A chemical inactivator is located at the dendrite of this postsynaptic neuron to counteract the impulse generated by the neurotransmitter. Each neurotransmitter has a specific inactivator. For example, cholinesterase is the inactivator for acetylcholine, a neurotransmitter found in many of the body's muscular junctions. The inactivator stops the continuous transmission of the impulse unless a new impulse is generated by a neurotransmitter at the first neuron. But let us look at this process more closely to further understand the transmission of a nerve impulse.

Neurons produce transmitters either in their nerve endings or in their cell bodies. In the case of peptides, they are made in the cell bodies, and then transported to the nerve endings where they are then converted into transmitters. Some unmyelinated nerves, such as those in the sympathetic system, emit transmitters along the course of the nerve rather than the nerve ending. Transmitters pass through little bulges called varicosities located along the nerve fiber. These varicosities move along the nerve fiber, similar to when a ball moves through a stocking.

In other cases, transmitters are stored in tiny vesicles of the nerve endings. When an impulse or **action potential** affects these endings, calcium (Ca^+) ions pass into the endings from the extracellular fluid from the neighboring neutron. This connects the vesicle with the membrane of the nerve ending, and then the transmitter connects with the synaptic cleft. The upper layer of this four-layered membrane is presynaptic membrane, and the synaptic cleft is between this layer and the second layer. After the transmitter leaves the vesicle, it crosses the synaptic cleft and connects with the membrane or receptor site of the postsynaptic neuron. This receptor site is made up of various proteins, and those proteins that are specifically affected by a transmitter are known as receptors or targets of the transmitter. It is important to note that one transmitter can have different effects on different neurons. Acetylcholine can excite one neuron while inhibiting another. However, in vertebrates, GABA is always inhibitory, and glutamate is always excitatory (Figure 8.7).

Transmitters must be controlled. If their secretion was not stopped at some points, then rapid changes in excitation and inhibition would not occur, and all activity would be slowed down. Some synapses block impulse transmission rather than pass the impulse on to other synapses (known as an **excitatory synapse**). A synapse that inhibits impulse

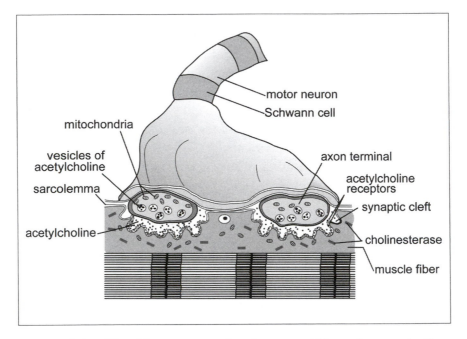

Figure 8.7　The Neuromuscular System. When the muscle fiber meets the motor neuron at the axon terminals, a movement or reaction results with the aid of neurotransmitters, including acetylcholine and its inactivator, cholinesterase. The sarcolemma contains receptors for the acetylcholine, and the mitochondria are where cell respiration takes place and energy is produced. Muscle contraction begins when the axon terminal receives a nerve impulse, which stimulates the release of acetylcholine. This release causes electrical changes—because of the movement of ions—at the sarcolemma. (Sandy Windelspecht/Ricochet Productions)

transmission is known as an **inhibitory synapse**. In this situation, a chemical inactivator located at the dendrite of the postsynaptic neuron inactivates the neuron, which ceases any impulse unless a new impulse from the first neuron releases more of the neurotransmitter.

　　Peptides have been discovered to be transmitters only in the past 20 to 30 years. One example is the parasympathetic nerves that control the salivary glands, which release a transmitter called VIP, also known as vasointestinal peptide. VIP increases the amount of acetylcholine released by nerve endings and also dilates blood vessels, which brings more blood to the glands so more saliva can be secreted. In addition, because of its ability to dilate blood vessels, it is used in the penis to achieve an erection. Often someone who is impotent has a deficiency of VIP in his genital organs.

Nerve Impulse

The body's nerve fibers system can be compared to a telegraph wire. Both are electric conduction systems designed to relay messages quickly over long distances. Both transmit messages in the form of pulses that are of a constant size and speed. Both the nerve fiber and the telegraph wire must be insulated for protection. If the wires or the fibers are damaged in any way, they cannot carry the impulses. But the nerve fiber is significantly more complicated than a telegraph wire. The nerve fiber generates the message itself along with transmitting it, whereas the telegraph wire is simply involved in the transmitting.

The body is composed of two kinds of fluid: **intracellular (ICF)** and **extracellular (ECF)**. ICF, the fluid within the cells, contains 65 percent of the body's total water. ECF, the fluid outside the cells, contains the remaining 35 percent. Both ICF and ECF contain electrolytes—chemicals that dissolve in water to become positive and negative ions. Positive ions are known as cations, while negative ions are called anions. ECF is comprised of a salt and chloride solution—NaCl—that breaks down to become sodium cations (Na^+) and chloride anions (Cl^-). ICF is composed of potassium cations (K^+). Both Na^+ and K^+ are essential for neurons to conduct impulses throughout the body. In order to understand the electrical changes, look at Figure 8.8.

When a neuron is not carrying an impulse, it is considered in a state of **polarization** (A). This means that Na^+ is more abundant outside the cell, and K^+ is more abundant inside the cell. In this state, the neuron has a positive charge on the outside of the cell membrane and a negative charge inside. When an axon receives a nerve impulse, it releases the neurotransmitter acetylcholine (ACh). ACh diffuses across the synapse and bonds to the ACh receptors, located on the **sarcolemma**. The ACh makes the sarcolemma permeable to the Na^+ ions. These Na^+ ions then rush into the cell membrane and the K^+ ions rush out of the cell. The neuron is then in a state of **depolarization** (B), and the charges acting on the membrane are

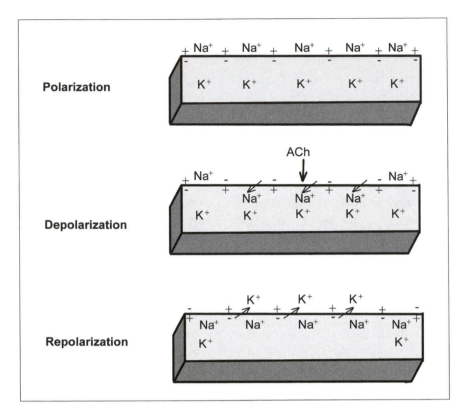

Figure 8.8   Muscular Electrical Charges. Muscle contractions are caused by electrical charges and the ion chemistry at the sarcolemma. Polarization is when the muscle is relaxed, depolarization is when the muscle is responding to the influence of acetylcholine, and repolarization is the process of the ion concentration returning to the polarization state. See also Table 1.1. (Sandy Windelspecht/Ricochet Productions)

reversed. Therefore, the inside of the cell has a positive charge whereas the outside has a negative charge. In addition, the cholinesterase inactivates the ACh. This is when the impulse is generated and transmitted through the body to prompt a reaction.

In order for the neuron to receive another impulse, it must be **repolarized** (C) and returned to the state it was in before it received the stimulus. Repolarization happens when the body's sodium pumps return the Na$^+$

TABLE 8.1
Electrical Charges

Polarization (resting potential)

• A positive (+) charge is outside the sarcolemma and a negative (–) charge is inside.

• Na⁺ ions are more abundant outside the cell. They diffuse inward until the sodium pump pushes them back outside the cell.

• K⁺ ions are more abundant inside the cell. They diffuse outward until the potassium pump pushes them back inside the cell.

Depolarization (action potential)

• ACh is released by the neuron when a nerve impulse is received at the axon terminal.

• ACh makes the sarcolemma permeable to Na⁺ ions.

• Na⁺ ions rush into the cell, causing a reversal of charges on the sarcolemma—now the outside is (–) and the inside is (+).

• The ACh is then deactivated by cholinesterase.

Repolarization

• Cell becomes permeable to K⁺ ions, which rush out of the cell.

• Charge restoration takes place—the outside is (+) and the inside is (–).

• The Na⁺ ions return outside via the sodium pumps, and the K⁺ ions return inside via the potassium pumps.

• Muscle fibers now are ready to respond to another nerve impulse received by the axon.

---

ions outside the cell membrane, and the potassium pumps return the K⁺ ions inside the membrane. The neuron is then once again in a state of polarization—the Na⁺ ions are more populous outside and the K⁺ ions are dominant inside. The neuron is now ready to respond to a stimulus and generate an impulse. Take a look at Table 8.1 for a recap of these processes. The polarization, depolarization, and repolarization cycle is similar to when a crowd of cheering fans at a stadium sporting event begin a "wave." One section will get up and throw their hands in the air, followed by the neighboring section, and so on, and a wave motion reverberates around the stadium. Just like the neuron's polarization behavior, the activity of each section influences the activity of the next section.

A neuron responds to a stimulus or action potential very quickly—in fact, it is measured in milliseconds. Each neuron has the ability to respond to hundreds of stimuli each second and to generate an electrical impulse.

Motor neurons are especially efficient because only their nodes of Ranvier depolarize, which is known as salutatory conduction. In these neurons, impulses travel extremely rapidly along the nodes of Ranvier. The neuron's myelin sheath also increases the rate at which impulses are generated.

Once the impulse is generated, neurons are able to transmit the information within milliseconds, even over great distances. For example, when a barefooted man or woman over six feet tall steps on a sharp tack, they will feel pain just as quickly as someone who is only five feet tall. The body can transmit these sensory impulses from the sole of a foot to the brain in under a second.

The nerve fibers that actually transmit information to the spinal cord and brain are called primary afferent axons. The speed at which these fibers can transmit messages depends on their thickness—the thicker the fiber, the faster information can travel. These axons are classified into four different groups (in order of decreasing size—thickest to thinnest):

- *A-alpha* carries information related to muscles

- *A-beta* carries information related to touch

- *A-delta* carries information related to pain and temperature

- *C* carries information related to pain, temperature, and itch

For example, if someone stubs her toe on a table, she initially feels the touch sensation when her toe collides with the table. The thick (and therefore fastest) A-beta nerve fiber carries this sensation from her toe to the brain. Because of the speed of this fiber, this sensation reaches the brain first. But shortly after, she feels pain. This is because the information related to pain is carried by the slower and thinner A-delta and C-nerve fiber.

The Spinal Cord

The central nervous system (CNS) is made up of two major organs: the brain and the **spinal cord**. The spinal cord connects the brain to the peripheral nervous system (PNS)—the nerves associated with the brain and spine. In order for information to travel between the brain and PNS, it must pass through the spinal cord.

Protected by a bony canal, the spinal cord extends down to the end of a column, which is made up of bones called vertebrae. This vertebral column grows at a faster rate than the nerve tissue of the spinal cord, so eventually the lower part of the canal grows longer than the cord. In an adult body, the spinal cord ends a short distance from the lowest rib (between the first and second vertebrae). An adult male's spinal cord will measure about 45 cm long, and an adult woman's will grow to about 43 cm long. The nerves that are below the spinal cord run in the vertebral canal—they are known as the cauda equina, which is Latin for "horse tail."

Function

The spinal cord has three main functions:

- *Direct reflexes*: Examples of a **reflex** are a knee jerk or pulling one's hand away from a hot surface. A **spinal reflex** only passes through the spinal cord and doesn't involve the brain.

- *Conduct sensory impulses*: Transmit sensory information from the afferent neurons to the brain through ascending nerve tracts.

- *Conduct motor impulses*: Transmit impulses from the brain through descending nerve tracts to the efferent neurons that communicate with the body's glands and muscles.

The **stretch reflex**—when a muscle is stretched and responds by contracting—is the most basic reflex arc in humans. This is basically a synapse between a sensory or afferent neuron and a motor or efferent neuron. Stretch reflexes can be induced by tapping many of the body's larger muscles, such as the triceps in the arm or the calf muscle in the leg. But most reflexes involve at least three neurons and numerous synapses. Reflex pathways or arcs will be examined more thoroughly later in this chapter.

Structure

Spinal Cord

The center of the spinal cord contains H-shaped material—this is the gray matter, which is made up of the cell bodies of the motor neurons and interneurons (Figure 8.9). Surrounding this gray matter and filling out the

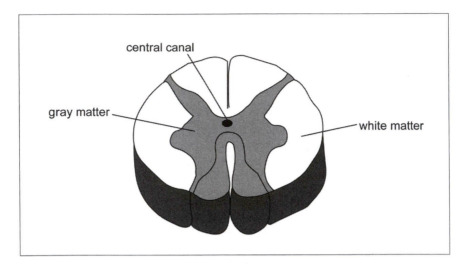

Figure 8.9   Another Cross-Section View of the Spinal Cord.
This cross section of the spinal cord shows the gray matter, which contains the cell bodies of the motor neurons and interneurons, and the white matter, which contains the interneurons' myelinated axons and dendrites. The central canal contains the cerebrospinal fluid. The gray matter coming out the top of the figure is the dorsal root, and the gray matter coming out the bottom is the ventral root. (Sandy Windelspecht/Ricochet Productions)

remainder of the spinal cord is the white matter—consisting of the interneurons' myelinated axons and dendrites. These are known as nerve fibers, which are bundled into nerve tracts based on what function they perform. The descending tracts carry motor impulses away from the brain, and the ascending tracts carry impulses to the brain. Also note the central canal—this contains the **cerebrospinal fluid** (CSF), which circulates in and around the brain and spinal cord.

Spinal Nerves
The spinal cord contains 31 pairs of spinal nerves (Figure 8.10). Because they are in pairs, they are bilateral, which means the nerves and nerve tracts occupy both sides of the spinal cord. Whereas the spinal cord is considered part of the central nervous system (CNS), the spinal nerves and nerve tracts are part of the peripheral nervous system (PNS).

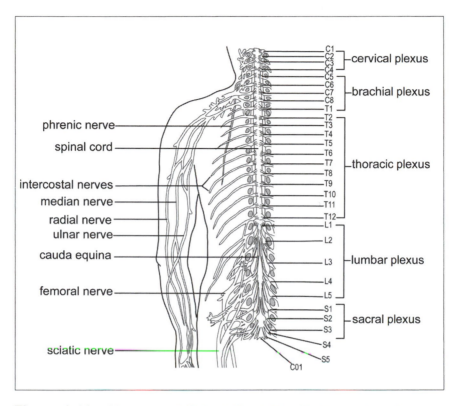

Labels: cervical plexus, brachial plexus, thoracic plexus, lumbar plexus, sacral plexus; phrenic nerve, spinal cord, intercostal nerves, median nerve, radial nerve, ulnar nerve, cauda equina, femoral nerve, sciatic nerve; C1 C2 C3 C4 C5 C6 C7 C8 T1 T2 T3 T4 T5 T6 T7 T8 T9 T10 T11 T12 L1 L2 L3 L4 L5 S1 S2 S3 S4 S5 C01

Figure 8.10   Nerves and Spinal Cord. The CNS's spinal cord consists of spinal nerves (on the left of the figure) and the spinal nerve plexus network (located on the right of the figure). The purpose of the nerve plexus is to combine neuron networks from various parts of the spinal cord to form nerve systems specific to parts of the body. For example, the arms' radial and ulnar nerves are part of the brachial plexus. (Sandy Windelspecht/Ricochet Productions)

As noted in Table 8.2, each pair of spinal nerves is numbered corresponding to the level of the cord and area of the vertebrae where it is attached. In addition, each pair is defined by a letter. For more information on spinal cord-related injuries, read Sidebar 8.1.

There are eight **cervical** pairs, 12 **thoracic** pairs, five **lumbar** pairs, five **sacral** pairs, and one tiny **coccygeal** pair (see Table 8.2). The first cervical vertebra is called the atlas, inspired by one of the heroes in Greek mythology. After losing an important battle, Atlas was turned to stone

TABLE 8.2
Spinal Nerves

Nerve group	Name	Corresponding location
CI–C8	Cervical pairs	Neck
TI–T12	Thoracic pairs	Ribs
LI–L5	Lumbar pairs	Large vertebrae in the small of the back
SI–S5	Sacral pairs	Base of the spine
COI	Coccygeal pair	Pelvic floor

and forced to carry the Earth and heavens on his shoulders. Therefore, this vertebra is called atlas because it carries the weight of the head.

Receptors located on the skin send information to the spinal cord through the spinal nerve. In referring to Figure 8.11, notice the **dorsal** and **ventral roots** protruding from the spinal cord. These roots attach the nerves to the cord. Dorsal refers to the back or posterior of the body, and ventral refers to the front of the body. The dorsal horns are the gray matter in the dorsal area—this is where sensory information is received through the dorsal root. The ventral horns are the gray matter in the ventral region, and they contain the motor neurons. The axons of these motor neurons leave the spinal cord, travel along the ventral roots, and head directly to the muscles.

Each dorsal root contains an enlarged area of gray matter, called the **dorsal root ganglion**. Because ganglion refers to any collection of nerve cell bodies outside the CNS, the dorsal root ganglion contains the cell bodies of the peripheral sensory neurons, such as the receptors on the skin. All the nerve fibers from all the sensory receptors throughout the body converge in the dorsal root ganglia.

The ventral roots contain motor nerve fibers, which connect to voluntary muscles, involuntary muscles, and glands. Cell bodies from these neurons are housed in the gray matter of the spinal cord. Both the dorsal and ventral roots meet in the spinal nerve—therefore creating a network of both sensory and motor nerves.

The groupings of nerves on the right side of the body are called **nerve plexuses**. A nerve plexus is a combination of neurons from various sections of the spinal cord that serve specific areas of the body (see Table 8.2).

SIDEBAR 8.1

Making Advances in Spinal Cord Injury Research

Treating spinal cord injuries and paralysis has presented some of the greatest challenges to medical researchers throughout history. Even as we understand more about the nervous system and how neurons behave, researchers could not uncover a way to re-grow injured nerves in the brain and spinal cord.

That was until two important scientific breakthroughs in the 1980s. First, experiments in rats showed that under certain laboratory conditions, many types of injured neurons in the central nervous system could be regenerated. Then, in the late 1980s, as scientists were focused on figuring out what was keeping axons from regrowing in people, they discovered important proteins that were inhibiting this growth. These proteins were produced by the oligodendrocytes. These are the cells that create the myelin sheath that covers nerves.

Both of these discoveries reinvigorated an area of research that had been considered beyond hope. Below are three areas where research is currently focused in treating spinal cord injuries:

Promoting Axon Growth: Researchers now know that it is not enough just to regrow a damaged axon. The axon needs to be positioned in the proper area of the spinal cord that will promote its growth and support its function. Even more challenging is that many regions of the adult spinal cord contain chemicals that inhibit the growth of a damaged axon, prompting it to retreat. Scientists are working on changing this environment in order to make it more hospitable to growing axons.

Enhancing Compensatory Growth of Uninjured Axons: Studies have shown that when damaged axons undergo treatment, the healthy neurons that surround the damage site begin to grow and support the recovering cells. Scientists are now using this knowledge to try to repair damaged nerve networks, particularly in patients who still have uninjured nerve networks. The hope is that these healthy nerves might be manipulated into taking over the function of the damaged nerves.

Preventing Scar Formation: Scar tissue that forms over the site of the injury can hamper repair. Scientists have discovered some molecular signals that scar tissue gives off that tries to block growth. Research is now focused on analyzing enzymes that will block these molecules and allow nerves to grow amidst scar tissue.

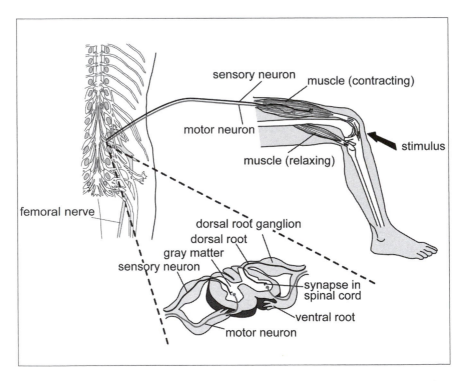

Figure 8.11   Spinal Roots of the Reflex Arc. A stimulus is applied to the knee, causing the sensory neuron to transmit the impulse to the spinal cord. The impulse is processed, and the instructions for a reaction are carried by the motor neuron, telling the relaxing muscle to contract. This reaction then causes the contracted muscle to relax. (Sandy Windelspecht/Ricochet Productions)

Differences among Spinal Cord Segments

All the pictures in Figure 8.11 represent a different segment of the spinal cord. Each picture is slightly different. The darker areas in each picture represent the gray matter, which is where the cell bodies of the nerve fibers are located, and the white matter is represented by the lighter areas that surround the gray matter. This is where the spinal cord's axons are located. In the cervical segment picture, there is a large amount of white matter. The cervical vertebrae fall right below the skull, at the top of the spinal cord.

Because of their location, there are many axons traveling up to the brain from all levels of the spinal cord. In addition, there are many axons traveling from the brain to the various segments on the spinal cord. In contrast, the sacral segment (which is the lowest, except for the coccygeal pair) has much less white matter. This is because fewer axons are traveling to and from the brain through this spinal cord segment. To summarize, the amount of white matter in relation to gray matter decreases as one moves down the spinal cord.

In addition to the differences in the amount of white matter, there are also differences in the size of the ventral horn, depending on the level of the spinal cord. Motor neurons are abundant and large in the segments controlling limbs, which are the lower cervical (C5–C8), lumbar, and sacral sections. In addition, the thoracic level features an extra cell column called the intermediate or **intermediolateral cell column**. This cell column is the location of all presynaptic sympathetic nerve cell bodies.

Reflexes

As mentioned earlier in this chapter, a reflex refers to when an incoming signal is processed by the CNS and then reflected to the motor nerve fibers, which then generate movement. But the action is generated from the spinal cord, and the brain is not directly involved.

The path that a nerve impulse follows after a signal is processed is known as the reflex arc. Five essential elements are involved in the reflex arc:

- Receptors detect the incoming stimulus and generate an impulse.

- Impulses are transmitted from receptors to the CNS through sensory neurons.

- The CNS houses the synapses where the impulse travels through.

- Impulses are transmitted from the CNS to the effector by motor neurons.

- The effector performs its distinctive action.

An appropriate example of a reflex arc is the patellar reflex, also known as the knee-jerk reflex. This is often an initial clinical test to determine if there is any neurological damage in the CNS. Doctors will hit the patellar tendon (right below the knee) with a rubber mallet to ensure a patient's nervous system is working correctly. If the knee quickly rises in response to the stimulus, then the CNS is functioning properly. Any problems with this response might indicate trouble in the thigh muscle or spinal cord. When the leg is raised, the muscle stretches and contracts; this is known as the stretch reflex.

In order to understand the patellar reflex, it is helpful to look at the different elements, which collectively happen in under one second. When the stimulus, or rubber mallet, hits the patellar tendon, the **stretch receptors** detect that the tendon is stretching. These receptors produce impulses that are carried along sensory neurons to the spinal cord. In the spinal cord, the sensory neurons synapse with the motor neurons, which transmit an impulse to the motor neurons in the femoral nerve. These neurons in the femoral nerve then transport impulses back to the quadriceps femoris (known as the effector), which contracts and then causes the lower leg to extend.

A closer look at a reflex, such as the patellar or stretch reflex, reveals the important role of the **muscle spindle**, which is a small group of muscle fibers wrapped in connective tissue that separates it from the rest of the muscle. Connected to an afferent neuron, the muscle spindle actually detects the stretch in the muscle.

Another important spinal cord reflex is the flexor, or withdrawal reflex. Once again, the flex reflex is automatic, and the brain is not directly involved in any decision making. This is when the stimulus is potentially harmful, such as touching a hot cooking pan or a sharp needle. The response is to pull one's hand or finger away. Similar to the patellar reflex, the sensory neurons transmit information to the spinal cord, and then the motor neurons tell the specific muscle to contract.

The Brain

Along with the spinal cord, the **brain** is the other major organ that makes up the central nervous system (CNS). Weighing approximately 3 pounds, or a little over 1 kilogram, in an average adult, the brain consists of over 100 billion neurons and trillions of **glia**, or support cells. The brain is covered by fluid, membranes, and bones. Housed in the **skull**, the brain is enclosed by a total of 14 bones, eight of which make up the **cranium** (the remaining six bones enclose various other parts of the brain). The skull is important because it protects the brain, and it will be explained more thoroughly later on in this chapter.

The brain is made up of many major parts that function as an integrated unit (Figure 8.12). The **brain stem** is the first major part. This consists of the medulla, pons, and midbrain, which is located just above the medulla. The remaining major parts are the cerebellum, the hypothalamus, the thalamus, and the cerebrum. Although each part is explored individually in the

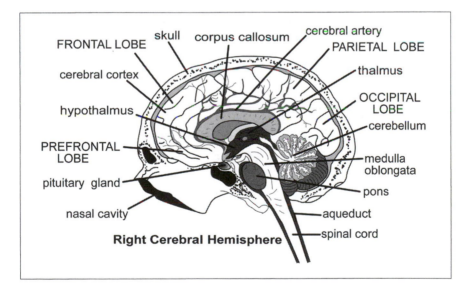

Figure 8.12 The Right Hemisphere of the Brain, including the Lobes. This hemisphere of the brain is associated with space-related perceptions, facial recognition, visual images, and music. (Sandy Windelspecht/ Ricochet Productions)

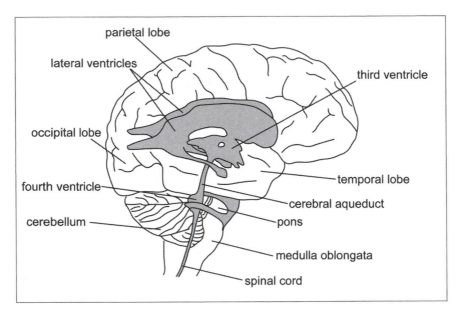

Figure 8.13 The Ventricles in the Brain. The brain's four ventricles are cavities that contain the choroids plexus, which is important in forming the CNS's tissue fluid (called the cerebrospinal fluid). (Sandy Windelspecht/Ricochet Productions)

following sections, it is important to remember that each part is intercon-nected and works with each other.

In addition to these major parts, the brain consists of four cavities, or ventricles (Figure 8.13). There are two lateral ventricles, as well as a third and fourth ventricle. Within each ventricle is a **choroid plexus**, a capillary network that forms cerebrospinal fluid (CSF) using blood plasma. This tissue fluid circulates in and around the brain and spinal cord, and will be covered in more detail later in the section about blood circulation in the brain.

Medulla

The **medulla** is anterior to the cerebellum, and extends between the spinal cord and the pons. Because the medulla contains myelinated nerve fibers and gray matter (or collections of cell bodies), this part of the brain is active in regulating the respiratory and cardiac activities of the body. This means regulating the body's heart rate, regulating the blood pressure by controlling the diameters (or width) of the blood vessels, and maintaining respiratory centers that control breathing. In addition, the medulla contains reflex functions that control coughing, sneezing, swallowing, and vomiting.

Pons

Also anterior to the cerebellum, the **pons** bulge out from the top of the medulla. The pons are also composed mainly of myelinated nerve fibers, which connect the two halves of the cerebellum with the brain stem in addition to the cerebrum above and the spinal cord below. This area of the brain is a vital link between the cerebellum and the rest of the nervous system. Nerve fibers in the pons carry messages to and from areas below it and above it. The pons contain two respiratory centers that work with the medulla to establish a normal breathing rhythm.

Midbrain

The **midbrain** extends from the pons and encloses the **cerebral aqueduct**, which is a tunnel that joins the third and fourth ventricles. The upper part of the midbrain is composed of four rounded masses of gray matter. Visual and auditory reflexes—or eye and ear reflexes—are housed in these masses. For example, when a bee comes buzzing toward someone's nose, he moves his head away. This is a visual reflex, because it involves the coordinated efforts of his eyeballs. When someone whispers near a person's ear, her first instinct will probably be to move closer to that whisper in order to hear better, which is an example of an auditory reflex. Righting reflexes—which ensure the head is upright and balanced—are also contained in the midbrain.

Cerebellum

The **cerebellum** is separated from the medulla by the fourth ventricle and cerebral aqueduct and is located below the **occipital lobes** of the cerebrum. As stated earlier, the cerebellum is connected to the brain stem, cerebrum, and spinal cord by the pons. This part of the brain controls movement, which includes coordination, regulation of muscle tone, and maintaining posture and equilibrium in the body. Equilibrium is actually controlled by receptors located in the inner ear. It is important to note that these are all involuntary functions. Because the cerebellum works below conscious thought, the conscious brain is able to function without being overwhelmed. For example, if someone drops a pen while writing, the cerebellum coordinates the impulses that tell her arm, hand, and fingers to pick up that pen. This all happens unconsciously so the brain can focus on tasks that need conscious effort, such as reading or writing.

Hypothalamus

Located at the base of the brain (above the pituitary gland and below the thalamus), the **hypothalamus** acts as the junction between the nervous system and the **endocrine system**. The hypothalamus is extremely small—it comprises approximately only 1/300 of the total brain weight. One of its most important jobs is to integrate all the various functions of the autonomic nervous system, which maintains the behavior of organs such as the heart, blood vessels, and intestines. But the hypothalamus has many other important and diverse jobs:

- *The body's thermostat*: The hypothalamus senses changes in body temperature and then transmits information so the body can adjust accordingly. For example, the hypothalamus will detect an increase in body temperature, which means the body is too hot. A signal will be sent to the skin's capillaries, telling them to expand to allow the blood to cool faster. Responses such as shivering or sweating are also prompted by the hypothalamus.

- *Oxytocin and antidiuretic hormone (ADH) production*: ADH helps to maintain the body's blood volume by enabling the kidneys to reabsorb water back to the blood. Oxytocin is important for women when they are in labor—it causes contractions that bring about delivery.

- *Stimulating the anterior pituitary gland*: The hypothalamus produces **growth hormone-releasing hormone** (GHRH), which stimulates the anterior pituitary gland to secrete **growth hormone** (GH). GH helps to encourage body growth throughout childhood—especially in bone and muscle development. But GH is also important in adults because it processes fats for energy production, increases the rate of cell division and protein synthesis, and speeds up the transportation of amino acids to cells, which enables protein production in the body.

- *Producing hunger sensations*: One of the hypothalamus's important jobs is to sense changes in blood nutrient levels. When these levels are low, this means that a person needs to eat to replenish nutritional resources. The hypothalamus activates this hunger sensation, so people eat, and blood nutrient levels are raised. This creates the sensation of fullness, so that a person stops eating.

- *Stimulating visceral reactions in emotional situations*: When one is feeling intense emotions, such as anger or embarrassment, the hypothalamus detects a change in the emotional state and prompts a response by the autonomic nervous system. People cannot control these visceral responses, and scientific experts still do not fully understand the biological and neurological bases for emotional reactions.

- *Maintaining circadian rhythms*: The hypothalamus regulates body rhythms, sleep cycles, and accompanying changes in mood and mental alertness. Our **circadian rhythm** and biological clock ensure that people are awake and alert during the day and tired at night. When people sleep, the hypothalamic biological clock is reset, and they are able to be awake for the day. If someone gets too little sleep, his clock might not be completely reset, and he will feel tired the next day.

Thalamus

The third ventricle passes through both the hypothalamus and **thalamus**. Located above the hypothalamus, the thalamus focuses **on sensations**. Almost all sensory impulses travel through the thalamus. With the exception of smell, the sensory impulses initially enter the brain through neurons in the thalamus. Eventually, these impulses are transmitted to the cerebrum, where they are processed (which eventually leads to perception), but initially, the thalamus categorizes and integrates these impulses. For example, when someone holds a glass of ice water, her body feels impulses related to coolness and the feel of the glass, including its texture and shape, which is perceived by sensory receptors in her muscles. She does not feel these sensations separately because the thalamus integrates them before sending them to the cerebrum, where they are interpreted and felt.

The thalamus can also block minor sensations, causing one not to be distracted while intensely focused on a particular task. For example, when someone is engrossed in a good book or television program, he might not notice someone coming into the room or speaking to him. The thalamus allows the cerebrum to focus on that book or television program by suppressing these distracting sensations.

Cerebrum

The largest part of the brain, the **cerebrum** is divided into two halves called hemispheres, which are separated by a deep groove or longitudinal fissure. The hemispheres are also connected by a bundle of nerves called the **corpus callosum**, found at the base of the longitudinal fissure. A band of 200 million neurons, the corpus callosum allows the right and left hemispheres to communicate with each other. The brain stem connects the cerebrum with the spinal cord. As stated earlier, it is also the general term for the area between the thalamus and the spinal cord, which includes the medulla and pons.

The outer layer of the cerebrum is called the **cerebral cortex**, which is a sheet of gray matter tissue, about 2 to 6 mm in thickness. The word "cortex" comes from the Latin word for "bark." The cerebral cortex is similar to the bark of a tree—it serves the same protective function. The gray matter is composed of the cell bodies of neurons, which carry out the many functions of the cerebrum. White matter is also located inside of the gray matter. Consisting of myelinated axons and dendrites, the white matter connects the cerebrum's lobes to one another and to other parts of the brain (the brain's lobes will be discussed shortly).

The cerebral cortex is folded many, many times in the brain. In humans, the cerebral cortex looks like it has many bumps and grooves. These folds or bumps are known as convolutions or **gyri** (plural of gyrus), and the grooves between them are called fissures or **sulci** (plural of sulcus). Because of this extensive folding, millions and millions of neurons are located on the cerebral cortex. The degree or extent of folding corresponds with the brain's capabilities. In an animal such as a cat or a dog, the cerebral cortex does not have nearly the amount of folding as in a human; therefore, animals cannot do many things that humans do—such as read, speak, or talk.

Lobes

Certain areas of the brain are associated with specific functions. In addition, the cerebral cortex is divided into lobes, which are also associated with certain brain activities. Each hemisphere is separated into a frontal lobe, parietal lobe, temporal lobe, and occipital lobe (Figure 8.14). Table 8.3 shows a summary of each lobe's functions.

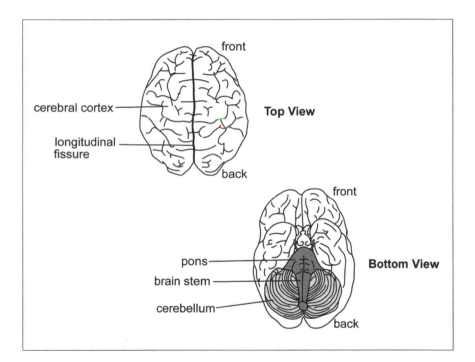

Figure 8.14 Views of the Brain. The longitudinal fissure separates the brain's left and right hemispheres. The pons, which are visible in the bottom view, help to regulate breathing, and the cerebellum regulates movement. The brain stem contains the pons, along with the medulla and midbrain. (Sandy Windelspecht/Ricochet Productions)

TABLE 8.3
Lobes of the Brain

Frontal lobe	Motor areas: Control movement
Parietal lobe	Sensory areas: Perceive touch, pressure, temperature, and pain
Temporal lobe	Auditory and olfactory areas: Hearing and smelling; speech areas and hippocampus
Occipital lobe	Visual areas: Vision, spatial distances

The **frontal lobes** contain the motor areas of the brain—here, impulses are generated for voluntary motor activity and movement. The left motor area (in the left brain hemisphere) controls movement on the right side of the body, and the right motor area (on the right brain hemisphere) controls movement on the left side of the body (Figure 8.15). When someone has a

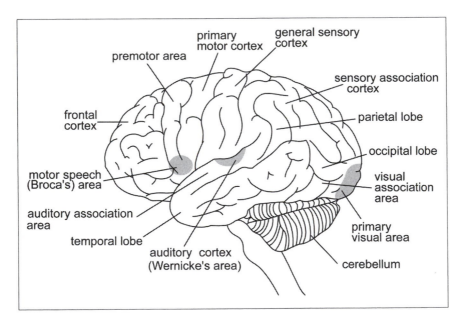

Figure 8.15 Functional Areas of the Brain. These areas work to regulate speech, cognition, sensory association, motor abilities, and visual processes, among others. (Sandy Windelspecht/Ricochet Productions)

stroke, also known as a cerebrovascular accident, either (or both) of their motor areas might be damaged, causing muscular paralysis. If the right motor area is damaged, then the left side of the body might be paralyzed. In addition to movement, the frontal lobe is involved in reasoning, planning, emotions, and problem solving. The frontal lobe in the human brain is relatively larger than in any other organism.

General sensory areas are located in the **parietal lobes**. This area focuses on receiving impulses relating to touch, pressure, temperature, and pain. Like the motor areas, the left area of the brain works with the right area of the body and vice versa. Skin receptors transmit impulses to the parietal lobes, where they are felt and interpreted. This area of the brain is also involved with stretch receptors in muscles and with taste—taste receptors, or taste buds, send their impulses to the taste areas of the parietal lobes. Taste areas overlap the parietal and temporal lobes.

Auditory (hearing) areas and olfactory (smelling) areas are located in the **temporal lobes**. Receptors for hearing are located in the inner ear, and olfactory receptors are located in the nasal cavity. In addition, temporal and parietal lobes in the left hemisphere contain speech areas that are involved with the actual thinking that precedes speech. Many people explain away an embarrassing slip of the tongue by exclaiming, "I spoke without thinking," but in reality, this is impossible.

The hippocampus, located in the temporal lobe and on the floor of the lateral ventricle, is involved with memory. Although little is known about how the brain actually stores and processes memories, it is believed that the hippocampus will collect information from many areas of the cerebral cortex, such as people's names or places where one has visited. If someone's hippocampus is damaged, they can only form memories that last a few seconds. This is evident in someone who suffers from Alzheimer's disease. Brain neurons are destroyed with this disease, followed by a loss of memory and the individual's personality. In addition to the memory functions, the hippocampus is also part of a group of structures known as the **limbic system**, which is important for controlling emotional responses, such as laughing and crying.

The last lobe area of the brain, the occipital lobe, is where the visual centers are located. Visual impulses received by the retinas in the eyes travel along the optic nerves to this area of the brain, where the brain processes and interprets what has been seen. Spatial relationships—such as judging distance and viewing in three dimensions—are processed in the occipital lobe.

Many areas in the cerebral cortex are not involved with movement or sensations. These are called association areas, and are believed to give people individuality and personality, including a sense of humor and the ability to learn and use reason and logic. See Table 8.4 for more details on the association areas of the cerebral cortex.

Two remaining important parts of the brain are the **basal ganglia** and the **corpus callosum**, which was mentioned earlier in this chapter. Grouped masses of gray matter within the white matter of the cerebral hemispheres, the basal ganglia regulate certain subconscious aspects of voluntary movement. Examples include movements such as hand gestures while talking or arm movements from front to back when walking. Parkinson's disease involves an impairment of the basal ganglia.

TABLE 8.4
The Cerebral Cortex

Cortex area	Function
Prefrontal	Emotion, problem solving, complex thought
Motor association	Complex movement
Primary motor	Stimulates voluntary movement
Primary somatosensory	Receives tactile (touch) sensory information
Sensory association	Processes multisensory information
Visual association	Processes complex visual information
Visual	Receives simple visual information
Wernicke's area	Comprehends language
Auditory association	Processes complex auditory information
Auditory	Processes sound qualities, such as loudness or softness
Speech center, or Broca's area	Produces and articulates speech

This part of the cerebrum is in charge of many high-level functions, including language and learning, although language is managed in the left cerebral hemisphere.

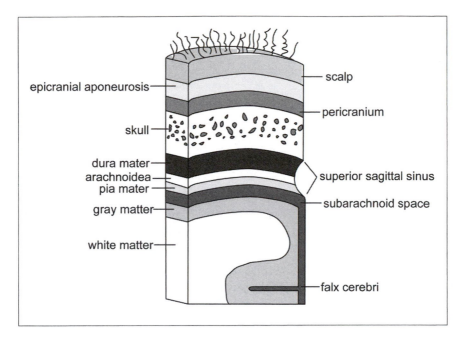

Figure 8.16   Cross-Section of the Skull and Brain. The scalp, pericranium, and skull all form the brain's coverings. (Sandy Windelspecht/ Ricochet Productions)

The corpus callosum is a band of nerve fibers that connects the left and right hemispheres of the brain and allows them to communicate and know each other's behavior (Figure 8.16). This creates a sort of "division of labor" in the brain, because each hemisphere performs functions the other does not. For example, the left hemisphere contains the body's speech centers. Because the right hemisphere does not, it has to be able to communicate with the left hemisphere through the corpus callosum to know what these centers are "talking about." See Table 8.5 for a summary of each hemisphere's functions.

Meninges

Because the brain is such an important organ, it needs a lot of protection. The initial layer of protection is the scalp, or skin layer, which contains the hair follicles. Right beneath the scalp is the skull, which is followed by

TABLE 8.5
Hemispheres of the Brain

Left hemisphere	Right hemisphere
Language	Spatial abilities and perceptions
Math	Facial recognition
Logic	Visual imagery, music

In general, each hemisphere of the brain is dominant for certain behaviors, although the information is shared through the corpus callosum.

three layers of connective tissues call **meninges**. Meningitis—a nervous system disease involving the brain—is an infection of the meninges.

The outermost layer of the meninges is made up of thick fibrous tissue called the **dura mater**, which lines the skull. Because of its thickness, the dura mater keeps the brain from moving too much in the skull, which could cause blood vessels to stretch and tear. The next layer is the **arachnoid membrane**. Arachnids are spiders, so arachnoid is an appropriate name for this layer because it is made up of web-like strands of connective tissue. Finally, the innermost layer is called the **pia mater**, which is closest to the brain and spinal cord. An easy way to remember these layers is, "The meninges PAD the brain"—"P," pia mater; "A," arachnoid membrane; and "D," dura mater.

Between the arachnoid and the pia mater is the subarachnoid space, which contains the clear, colorless liquid called cerebrospinal fluid, or CSF.

CSF and the Ventricular System

Remember that the brain contains four ventricles, or cavities—the two lateral ventricles, and a third and fourth ventricle. Within each ventricle, there is a choroid plexus, a capillary network where CSF is formed from cellular secretions and filtration of the blood, also known as blood plasma.

CSF flows from the lateral and third ventricles through the fourth ventricle. Then it continues to the central canal of the spinal cord, then to the cranial and spinal subarachnoid spaces.

CSF is reabsorbed into the blood as more is produced. It is through this continuous process that it flows in and around the CNS. When CSF is in the cranial subarachnoid spaces, it is reabsorbed through the **arachnoid villi** into large veins within the dura mater called **cranial venous sinuses**. After this reabsorption, the CSF becomes blood plasma again. The rate of reabsorption usually equals the rate of production. Total daily production of CSF is normally about 13.53–16.91 fluid ounces (400–500 milliliters) and the total volume of CSF remains around 4.28–5.07 fluid ounces (125–150 milliliters).

The CSF is vital to the brain and spinal cord, and has many functions:

- *Protection*: CSF cushions the brain, especially if the brain is impacted by a blow to the head. However, the CSF can only provide so much protection—sharp or heavy blows will injure the brain.

- *Buoyancy*: Pressure at the brain's base is reduced because it is immersed in CSF, reducing the net weight of the brain from about 52.91 ounces to 1.76 ounces (1,500 grams to 50 grams).

- *Waste product elimination*: The CSF takes harmful substances and toxins away from the brain.

- *Hormone transportation*: The CSF transports hormones to all areas of the brain.

Blood Supply

Blood brings necessities such as oxygen, carbohydrates, amino acids, fats, hormones, and vitamins to the brain. In addition, blood removes carbon dioxide, ammonia, and lactate from the brain (Figure 8.17). The brain occupies about 2 percent of the total body weight in humans, but it receives about 15 percent of the blood supply. The brain has priority over all other organs in the human body for blood, because it needs blood to survive. Cells in the brain will die without oxygen, and it is the blood vessels within and on the surface of the brain that transport food and oxygen. Blood vessels bring their goods to the brain holes in the skull called foramina.

Memory and Learning

Memory is the mental ability to remember ideas and experiences. As stated earlier, scientists believe the hippocampus is responsible for

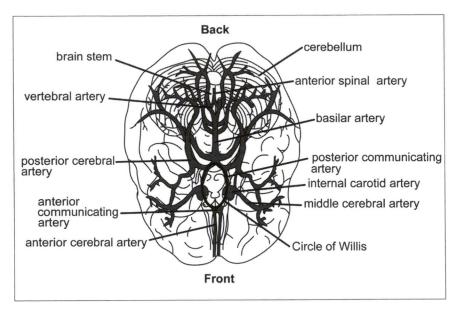

Figure 8.17   Arterial Networks of the Brian. The posterior communicating artery, internal carotid artery, and middle cerebral artery form the circle of Willis, which provides the brain with blood if one of the internal carotid arteries or one of the vertebral arteries gets blocked. (Sandy Windelspecht/Ricochet Productions)

forming, sorting, and then storing memories. The hippocampus is also responsible for connecting new memories with related ones, thus giving them meaning. For example, one might remember the first baseball game she played by associating it with the smell of the grass on the field, the feel of the uniform she was wearing, and the sound of the crowd cheering for her. Each of these is an individual memory that relates to an overall recollection of that particular experience.

The initial stage of memory processing involves recognizing visual and auditory (sights and sounds) sensory signals that are stored for only a minute or fractions of a second. This short-term memory retains these bits of information only for a short time, and they are lost if not reinforced or repeated. When stored information is recalled at a later time, this is called long-term memory. Memories become more ingrained in the mind the more often they are repeated. Therefore, a short-term memory can become a long-term memory. In addition, the memory becomes stronger and more impressed in the brain the more often it is recalled—this is referred to as memorization.

For example, say someone is studying a history book on the Revolutionary War for an upcoming test. The more times he reads this book and processes the information, the easier it will be for him to recall what he reads when it comes time to take the test. Certain facts, such as dates of battles, will be imprinted on his memory; the deeper that imprint, the easier and quicker it will be for him to remember these facts. Studies have shown that someone who is wide awake and mentally alert memorizes far better than someone who is tired.

Additional studies have shown that the brain is able to organize new information where similar information is stored. If, next month, that same person reads a book about George Washington becoming the first president of the United States, his memory will sort and file these new facts in the same area where it stored what he learned about the Revolutionary War. The memory will relate the new information on George Washington to what he learned about him during the Revolutionary War.

Peripheral and Autonomic Nervous System

As mentioned earlier, the human body's nervous system is separated into two divisions: the central nervous system (CNS) and the peripheral nervous system (PNS). The previous sections of this chapter explored the brain and spinal cord, the two primary components of the CNS. Through cranial and spinal nerves, the PNS transmits information to the CNS, where the brain processes it and responses are initiated.

The internal or visceral organs in the body, such as the heart and lungs, have nerve fibers and nerve endings that conduct messages to the brain and spinal cord. However, people are not aware that many of these messages reach their brain, although they do know that they happen or they would not be functioning. In other words, these impulses never reach their consciousness. These impulses are processed or translated into reflex responses without ever reaching the conscious areas of the brain. For example, people do not notice when their blood vessels expand or their heart rates increase; they happen involuntarily.

These efferent or visceral neurons are grouped together in the autonomic nervous system (ANS), which falls under the direction of the PNS. This is where visceral neurons, which are neurons associated with the body's internal organs, relay information to the glands in addition to the smooth and cardiac muscles. Through nerve networks, the ANS facilitates communication between sensory impulses from the blood vessels, heart, and organs located in the chest, abdomen, and pelvis to various parts of the brain (especially the medulla, pons, and hypothalamus). Bypassing the consciousness, these impulses elicit mostly automatic reflex responses in the heart, vascular system, and bodily organs that control temperature, posture, food intake, and reactions to stressful feelings (such as anger or fear), among other processes.

Cranial Nerves

The ANS consists of 12 pairs of **cranial nerves**, which originate in the midbrain, the pons, and the medulla of the brain stem (Figure 8.18). The pairs are numbered using Roman numerals—beginning in the front and ending in the back—that are based on the nerves' connection with the brain.

These cranial nerves are divided into sensory, motor, or mixed nerves depending on their function. Some of these nerves transport information from the sensory organs to the brain, and other cranial nerves work to control muscles. Some other cranial nerves control glands and internal organs, such as the ear and lung. Mixed nerves, which contain at least one sensory (or afferent) and one motor (or efferent) nerve, originate in more than one nucleus. In some cases, a single nucleus can produce more than one nerve. One example is the sense of taste, which comes from one nucleus even though its function is spread across two nerves.

Based on the sensory or motor functions of the cranial nerves, each pair is further defined by one of the following four categories:

1. *Special sensory impulses*: Senses relating to smelling, tasting, seeing, and hearing.

2. *General sensory impulses*: Senses relating to pain, touch, temperature, deep muscle sense, pressure, and vibrations.

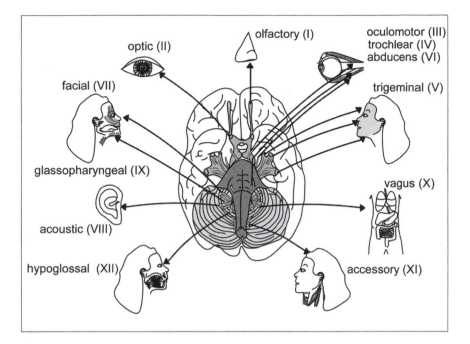

Figure 8.18   The Cranial Nerve Network. The distribution of the brain's cranial nerve network. (Sandy Windelspecht/Ricochet Productions)

3. *Somatic motor impulses*: These relate to voluntary control of the skeletal muscles.

4. *Visceral motor impulses*: These relate to involuntary reactions of glands and involuntary muscles such as the cardiac muscle.

Nerve pairs I, II, and VIII contain only sensory fibers, whereas III, IV, VI, XI, and XII contain mostly motor fibers. Pairs numbered V, VII, IX, and X contain mixed nerves. Following is an examination of the cranial nerves based on their functions.

Optical Sensations and Eye Muscles

Nerves III, IV, and VI are motor nerves that help the external eye muscles to operate. The **oculomotor nerve**, or Nerve III, controls blinking and pupil dilation. When one's pupils are dilated at the optometrist's office, the person receives eye drops, which are actually an acetylcholine blocker that inhibits the body's parasympathetic system, which will be explained later in this chapter. Once this happens, the pupils dilate, or expand, and the doctor can look inside the eye lens. However, the patient will not be able to read or focus on close objects because the parasympathetic system is being restrained.

The sensory aspect of this category involves Nerve II, the optic nerve, which transmits visual impulses from the eye to the brain. The other two cranial nerves in this category are motor nerves—the trochlear (Nerve IV) and the abducens (Nerve VI) control eyeball movement.

Face and Mouth Sensations

Nerve V is a mixed nerve that controls all sensations and movements from the face and mouth. For example, this nerve, also called the **trigeminal nerve**, controls chewing by carrying motor fibers from the face and mouth to the **mastication muscles**. Another example is getting a hard knot in the cheek when clenching the teeth. This happens when the trigeminal nerve stimulates the **masseter muscle**.

The sensory functions of this nerve include carrying impulses such as pain, touch, and temperature to the brain. For example, when someone is hit in the face, this nerve senses the onset of pain and changes in touch and transmits these impulses to the brain.

Another nerve in this category, Nerve VII, is simply known as the facial nerve. All of the muscles involved with facial expression are controlled by this nerve. However, the facial nerve is mixed because it also carries sensory impulses for taste from the tongue to the brain. Taste fibers, which originate from the taste buds, are predominantly located on the anterior two-thirds of the tongue. However, the touch and pain sensations related to the tongue come from Nerve V.

Also classified in this category is the **olfactory nerve**, or Nerve I. This sensory nerve gathers sensations relating to smell from the nasal mucosa (the membrane located in the nose's interior that produces mucus) to the brain.

Hearing and Balance

The **vestibulocochlear nerve**, or Nerve VIII, carries auditory or acoustic information relating to sound from the ear to the brain. In addition to hearing, this sensory nerve controls balance.

Throat and Salivary Glands

Nerve IX, or the **glossopharyngeal nerve**, is a mixed nerve that contains sensory fibers for the throat and for taste from the posterior one-third of the tongue. The tongue's movement is controlled by the **hypoglossal nerve**, or Nerve XII. Also located in this category are the motor nerves that control swallowing in the throat. In addition, this nerve contains sensory fibers that control secretions from the salivary gland in the throat. The **spinal accessory nerve**, or Nerve XI, also controls the throat muscles, in addition to the two major neck muscles.

Thoracic Nerve

Thoracic refers to the rib area of the spinal cord. Nerve X is the longest cranial nerve. Also known as the **vagus nerve**, it is a mixed nerve that controls most of the thoracic and abdominal organs, such as the glands, digestion (including the production of digestive juices), and the body's heart rate. In addition, the vagus nerve contains motor fibers that control the voicebox.

Autonomic Nervous System

An important part of the PNS, the ANS is made up of the motor elements of the cranial and spinal nerves. The ANS consists of visceral motor neurons that connect with smooth muscle, cardiac muscle, and glands. These areas are known as visceral effectors; they receive the impulses from the neurons through the nerve pathways and produce involuntary responses. For example, upon receiving impulses, the heart beats, muscles contract or relax, and glands secrete.

There are two divisions in the ANS: sympathetic and parasympathetic (Figure 8.19). In most cases, one will function in opposition to the other. The afferent nerves working for both divisions transmit impulses from sensory organs, smooth muscles, and the circulatory system, in addition to all the body's organs, to the vital centers of the brain. Then efferent impulses are conveyed from these centers to all parts of the body by way of the parasympathetic and sympathetic nerves, which will be discussed shortly.

Basically, the sympathetic division operates in stressful situations, and the parasympathetic controls the body in non-stressful situations. However, both divisions operate under the direction of the hypothalamus. As

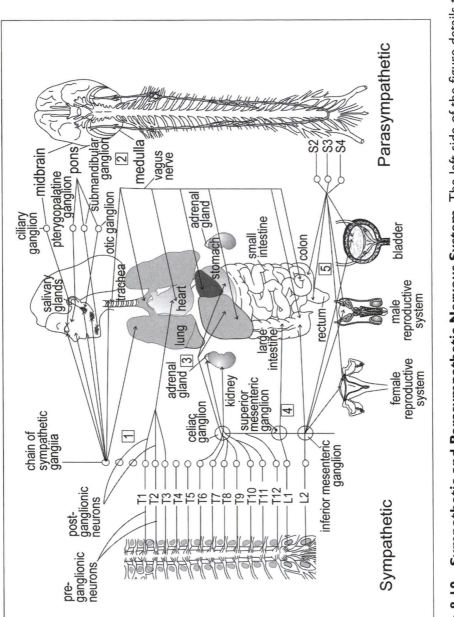

Figure 8.19   Sympathetic and Parasympathetic Nervous System. The left side of the figure details the sympathetic system, along with the preganglionic and postganglionic neurons and the chain of sympathetic ganglia that connects the nerves to their effectors. On the right side is the parasympathetic system, which has various ganglia that connect the nerves to effectors. (Sandy Windelspecht/Ricochet Productions)

mentioned earlier, the hypothalamus is the part of the brain located right below the thalamus and above the pituitary gland. The hypothalamus has many functions, including regulating body temperature and ensuring that the visceral effectors (smooth muscle, cardiac muscle, and glands) respond appropriately to the impulses transmitted by the visceral motor neurons.

It is important to point out the role of autonomic pathways in the PNS. Impulses travel from the CNS along the autonomic nerve pathway to visceral effectors. Along this nerve pathway are located two motor neurons that synapse outside the CNS in a ganglion, which is a group of neuron cell bodies. The first motor neuron is called the **preganglionic neuron**, which connects the CNS to the ganglion, and the second neuron is called the **postganglionic neuron**, which connects the ganglion to the visceral effector. The ganglion is made of cell bodies of the postganglionic neuron. Fibers of the preganglionic neuron are short and myelinated, which means parts are covered with a myelin sheath that provides electrical insulation and increases the speed at which an impulse is transmitted. In contrast, fibers of the postganglionic neuron are long and unmyelinated.

These ganglia are like relay stations in the body. An impulse or message is transferred at the synapse (located in the ganglion) from the preganglionic neuron to the postganglionic neuron and then to the muscle or gland. This is in contrast to the voluntary nervous system, where the motor nerve fiber functions by beginning at the spinal cord and extending to the skeletal muscle without a synapse. Through these processes of the autonomic pathways, stimuli is transmitted through the body, eliciting unconscious, automatic reflex responses such as the stomach and intestinal digestion, respiration rate and depth, pupil dilation, and blood rate regulation through the expansion or contraction of blood vessels.

Sympathetic Nervous System

The cell bodies of the sympathetic division's preganglionic neurons originate in the thoracic and some of the lumbar segments of the spinal cord, which are located at the small (or lower area) of the back. Because of this, the sympathetic division is often called the thoracolumbar division. The ganglia of this division are located in two chains right outside of the spinal column. The sympathetic ganglia connect with them on these two chains. The ganglia are home to the synapses between the preganglionic and postganglionic neurons. After receiving an impulse, a preganglionic neuron synapses with a postganglionic neuron to initiate a response in an effector. It is important to note, however, that one preganglionic neuron often synapses with many postganglionic neurons, and thus many effectors. This allows the simultaneous responses among many effectors—a vital function of the ANS.

When a person becomes angry or afraid, or falls under any kind of stress, the sympathetic division of the ANS becomes dominant in the body (this also includes during exercise). This was essential for the survival of humans' ancient or prehistoric ancestors. Their lifestyle included hunting live animals for food and protecting themselves and their families from enemies such as animals and other humans. This demanded intense physical activity and endurance. This division of the ANS is often referred to as the "fight-or-flight response," which was obviously an important inner navigating tool for avoiding danger and defending family and property.

Even thousands and thousands of years later, the human nervous system is not much different than that of its prehistoric forefathers. The fight-or-flight response still helps the body to determine the appropriate behavior when it feels afraid or anxious. Table 8.6 shows how some of the body's organs respond when in a stressful situation. The heart rate increases, and breathing becomes heavier because the bronchial muscles are contracting and the bronchioles are dilating, allowing for increased air intake. In other words, the breathing rate increases. A person might feel more powerful because the liver is changing glycogen to glucose, which supplies the body with more energy. Blood vessels associated with the visceral organs and the skin constrict, thus forcing more blood to vital organs such as the heart, muscles, and brain. But not all organs move at a faster rate. Stomach digestion is not important in stressful situations; therefore, secretion of digestive juices decreases along with peristalsis, the waves of muscle contractions that move food through the stomach muscle. All of these responses enabled prehistoric ancestors to stay and fight or run away from potential danger. People often find themselves in stressful situations that are not life-threatening, such as when they are studying for an important history exam or interviewing for a new job. But the body is prepared to react appropriately when circumstances escalate from everyday stress to life-threatening danger. See Sidebar 8.2

TABLE 8.6
Autonomic Nervous System: Visceral Effectors

Effector	What happens when the sympathetic division is activated?	What happens when the parasympathetic division is activated?
Eye muscle (also known as an iris)	Pupils dilate	Pupils contract
Salivary glands	Saliva production decreases	Saliva production increases
Nasal and oral mucus (mucosa)	Mucus production decreases	Mucus production increases
Heart	Heart rate increases	Heart rate decreases
Lungs	Bronchial muscle relaxes	Bronchial muscle contracts
Stomach	Peristalsis reduces	Peristalsis increases, gastric (or digestive) juices are secreted
Small intestine	Digestion processes slow down	Digestion processes increase
Large intestine	Movement and contractions slow down	Secretions increase, along with movement and contractions
Kidney	Urine secretion decreases	Urine secretion increases
Bladder	Organ wall relaxes and sphincter closes	Organ wall contracts and sphincter relaxes or opens
Liver	Glycogen changes to glucose	None
Sweat glands	Production of sweat increases	None
Skin and viscera blood vessels	Constrict	None
Skeletal muscle blood vessels	Dilate	None
Adrenal glands	Secretion of epinephrine and norepinephrine increases	None

to read how hormones enable the body to react appropriately in these situations.

An example of an everyday stress relieved by the sympathetic division is when the body temperature rises when one is sitting by the pool or on the beach. When someone is in a warm environment, this external stimuli can drain the body's heat reserves. Within the sympathetic division, thermal receptors send messages to the brain through the sympathetic nerve systems. One result of these messages is the expansion of cutaneous blood

---

SIDEBAR 8.2

How the Adrenal Medulla Helps Us Fight

The body's endocrine system controls glands that produce or secrete chemical messengers, known as hormones, into the bloodstream. One of these glands, which is stimulated by the ANS, is the adrenals. The adrenals are located right above the kidneys, and each is made of two small glands that operate independently. The inner area is called the medulla, and the outer area is called the cortex. The two main hormones released by the adrenal medulla, epinephrine (also known as adrenaline) and norepinephrine, are related to reactions to stressful situations also known as the "fight-or-flight response." Once these hormones are released, some of the effects include an increase in the heart rate, rate at which the body cells metabolize; dilation of the lungs' bronchioles, so the breathing rate can increase; and the conversion of glycogen (which is stored in the liver) to glucose. Through the blood, this glucose is sent to the voluntary muscles, enabling them to handle an increased workload.

---

vessels, which reside right below the skin's surface. The expansion or dilation of these blood vessels enables more blood to flow to the body's surface, where heat has been lost. This dilation also may cause oozing of certain fluids from the capillaries, causing dependent limbs to swell, which is why people sometimes swell in the heat. This increase in blood flow ensures that all necessary organs are receiving the proper amount of blood. Otherwise, the blood might concentrate in the lower limbs and not get to the brain, which leads to fainting spells.

If an organ receives both sympathetic and parasympathetic impulses, the responses are opposite. Note that some effectors can only receive sympathetic impulses (liver, sweat glands, many blood vessels, and adrenal glands). In these cases, an opposite response happens when there is a decrease in the sympathetic impulse.

Another important way the sympathetic division responds to excessive heat is by forcing the body to sweat. The brain's hypothalamus senses a great increase in temperature and conveys this information to the sweat glands via the sympathetic nerves, which causes one to sweat. Through the evaporation of sweat (sometimes aided by a cool breeze), the body cools down. Once again, this is all involuntary. However, one can cool the body by jumping in a pool or cool shower or even sitting in an air-conditioned room, which lowers the environmental temperature.

Parasympathetic Division

Another name for the parasympathetic division is the **craniosacral division**. In this division, the cell bodies of preganglionic neurons are located in the brain stem and sacral segments of the spinal cord. The axons of these preganglionic neurons are contained in cranial nerve pairs III, VII, and IX in addition to some sacral nerves. These axons extend to the parasympathetic ganglia, which are located in or extremely close to the visceral effector. The postganglionic cell bodies are actually located in the effector, and their very short axons connect with the cells of the effector.

Unlike in the sympathetic division, one preganglionic neuron synapses with only a few postganglionic neurons and then to only one effector. This aspect of the parasympathetic division enables single-organ responses, or very localized responses.

As stated earlier, this division dominates the body during non-stressful and relaxed situations, allowing several organ systems to function at a normal level and rate. For example, digestion is efficient through increased secretions and peristalsis. Urination and defecation occurs, and the heart rate will be at the normal resting rate.

Neurotransmitters in the ANS

As mentioned, there are two kinds of synapses in the ANS that bring about a reaction in a visceral effector: one between preganglionic and postganglionic neurons, and another between postganglionic neurons and the effectors. In order for nerve impulses to cross synapses, they need the help of neurotransmitters.

Acetylcholine is the neurotransmitter released by all preganglionic neurons in both the sympathetic and parasympathetic divisions. A chemical inactivator is located at the dendrite of each postsynaptic neuron to defuse the impulse generated by the neurotransmitter. This is because transmitters must be controlled and inactivated—if their secretion was not stopped at some point, then rapid changes in excitation and inhibition would not occur, and all activity would be slowed down. Acetylcholine's inactivator is cholintesterase, which is located in postganglionic neurons. However, in the parasympathetic division, postganglionic neurons also release acetylcholine right before they connect with their visceral effectors. In addition, most postganglionic neurons in the sympathetic division also release the neurotransmitter norepinephrine when they synapse with effector cells. Norepinephrine's inactivator is **catechol-O-methyl transferase**, or COMT.

The Senses

The central nervous system's five senses—seeing, hearing, touching, tasting, and smelling—allow the body to maintain homeostasis, which is when the internal environment is stabilized despite what constant changes are occurring in the external environment. The senses also protect people by providing information about what is going on inside and outside the body. For example, smelling and tasting might tell someone that something she is about to eat could be dangerous. Our touch sensation tells someone that a stove is too hot and will burn his skin on contact.

The information collected by the senses is transmitted through pathways and stimulates electrical nerve impulses. There are four important components of these sensory pathways:

- *Receptors*: Changes, or stimuli, are detected by receptors. All receptors respond to stimuli by generating electrical nerve impulses. However, depending on location, each receptor only responds to certain sensory changes. For example, receptors in the retina detect light rays, while nasal cavity receptors detect airborne chemicals. When the specific stimuli is detected, an impulse is generated.

- *Sensory neurons*: These neurons take the impulses produced by the receptors and transmit them to the central nervous system. Although the sensory neurons are located in the spinal and cranial nerves, each carries impulses from only one type of receptor. For example, separate networks of these neurons serve the eyes, nose, ears, skin, and mouth.

- *Sensory tracts*: Impulses are transmitted to a specific part in the brain through sensory tracts in white matter located in the brain or spinal cord. White matter is defined as nerve tissue composed of myelinated axons and dendrites.

- *Sensory area*: This is where the impulses, or sensations, are felt or perceived and interpreted. Located in the cerebral cortex, this area functions without a person's conscious awareness.

Breakdown of Sensations

Sensations have several important characteristics that enable people to feel, see, hear, smell, and taste. The first characteristic is **projection**. When a hand pets a furry cat, it seems like the sensation is located in the hand. However, receptors located in the hand collect information associated with the cat's fur and transmit it to the cerebral cortex or the brain, where it is interpreted as soft and fluffy. The brain projects what it feels to the hand. This aspect is evident in patients who have had a limb amputated. Often, patients say that even though their hand has been removed, they feel as if that hand is still there. Even though the hand's receptors have been removed with the severed limb, the nerve endings associated with those receptors still continue to generate impulses. The brain continues to behave as it did when the hand was still present. When the impulses from these severed nerve endings travel through sensory pathways to the brain, these impulses are interpreted and projected. The brain projects the sensation or feeling of the hand as still present. This feeling is known as **phantom pain**, and generally diminishes as the severed nerve endings heal.

Another important sensory characteristic is intensity, which is how strongly sensations are felt. A weak stimulus, such as a soft hum or dim light, will affect only a small number of receptors. However, a strong stimulus, such as a loud bang or bright light, will affect many more receptors, causing an increased amount of impulses to travel to the brain's sensory area. Based on the number of impulses received, the brain will respond accordingly. The more impulses received, the more intense the brain's sensory projection.

The brain's interpretations also allow it to contrast previous and current stimulations and allow for inflated or diminished sensations. For example, when someone takes a hot shower, the brain will compare the water temperature to those previously experienced. If the water is hotter than experienced before, the brain will most likely cause one to jump away from the water. But if the water is cooler than usual, the brain will tell one to make it hotter.

A third characteristic of the senses is **adaptation**, or when the body adjusts to a continuing stimulus. Receptors are always ready to detect changes to the body's external environment, but if the stimulus continues, it becomes less of a change. Therefore, the receptors will generate fewer impulses to the brain, which adapts itself to the stimulus. For example, many people wear jewelry on hands and arms, such as rings or watches. The presence of a ring is a continuous stimulus from the moment it is put on in the morning until it is taken off before bed. However, because it is on all day, the cutaneous or skin senses adapt to the presence of the ring, and the wearer becomes unaware that it is on his finger. Only when there is a change, such as when the ring comes off before bedtime, do the receptors detect a change.

After-image is the final characteristic of the senses. This is when a sensation remains in the conscious memory even after the stimulus has ceased. One example is a flashbulb from a camera, which often stays in the memory for a few minutes after a picture is taken. Because the flashbulb produces such a bright light, the receptors in the retina generate many impulses that are interpreted by the brain as an intense sensation. The sensation is so strong that it lasts a bit longer than the stimulus from which it was generated.

Cutaneous Senses

As mentioned earlier, **cutaneous senses** are those related to the skin (Figure 8.20). The cutaneous senses tell what is happening in and to the immediate external environment, including what is happening to the skin. For instance, an annoying mosquito bite on the knee often produces an itching sensation. This is actually a mild form of the pain sensation. The brain interprets sensory impulses in the parietal lobes. The largest part of this sensory area is reserved for the parts of the skin with the most receptors, which are the hands and face.

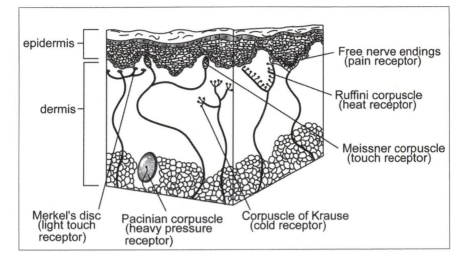

epidermis

dermis

Free nerve endings
(pain receptor)

Ruffini corpuscle
(heat receptor)

Meissner corpuscle
(touch receptor)

Merkel's disc
(light touch
receptor)

Pacinian corpuscle
(heavy pressure
receptor)

Corpuscle of Krause
(cold receptor)

Figure 8.20   The Cutaneous Senses. The cutaneous receptors of the skin include the Merkel's disc, the Pacinian corpuscle, the Ruffini corpuscle, and the free nerve endings. (Sandy Windelspecht/Ricochet Productions)

Sensations related to touch, pressure, pain, and temperature (heat and cold) are produced by receptors located in the skin's inner layer, or dermis. Pain receptors, also called free nerve endings, react to any intense stimulus. This means intense extremes of temperature applied to the skin, whether cold or hot, will be felt as pain. Encapsulated nerve endings are the receptors for the other cutaneous senses. This means that these nerve endings are surrounded by a cellular structure.

Taste and Smell Senses

Taste-specific receptors are located in taste buds, which are found in the **papillae** area of the tongue (Figure 8.21). In general, experts believe there are four types of taste receptors: sweet, sour, salty, and bitter. The papillae's taste receptors (or chemoreceptors) decipher chemicals from food that have dissolved in the mouth's solution, also known as saliva. A moist mouth full of saliva is necessary for taste distinction—if the mouth is dry, even the most flavorful food, such as a grilled steak, will have little taste.

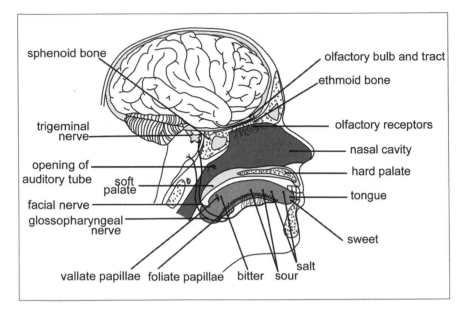

Figure 8.21 Smell and Taste Senses. This midsagittal section of the brain shows the smell and taste nerves and sensory components. (Sandy Windelspecht/Ricochet Productions)

Most foods contain a variety of the four general flavors and stimulate different combinations of receptors. For example, one bite of a peanut butter and jelly sandwich will stimulate both sweet (from the jelly) and salty (from the peanut butter) taste combinations. The smell of food also contributes to food perception.

Impulses relating to taste sensations are transmitted through the facial and glossopharyngeal nerves, which are the seventh and ninth cranial nerves, to the parietal-temporal cortex of the brain where the taste areas are located. Food scientists have found evidence of a genetic link in some taste preferences. For example, people who have an above-average number of taste buds tend to find broccoli bitter and unpleasant, whereas those with an average number of taste buds might like this vegetable's taste.

Olfaction, or the sense of smell, functions through chemoreceptors located in the upper nasal cavities. These receptors detect vaporized chemicals and then generate impulses that travel through the first cranial or olfactory nerves on the ethmoid bone. From there, these impulses move on to olfactory bulbs and then on to the olfactory areas of the brain's temporal lobes. Scientists believe there are at least 1,000 different smells.

In comparison to other animals, humans have a poor sense of smell. For instance, dogs have a far greater sense of smell—they are believed to smell 200 times better than humans. The olfactory sense has a rapid adaptation rate, which is why pleasant smells tend to be acute and sharp at first but then quickly fade away. As mentioned above, the taste sense is greatly influenced by smell, which is why food can lose its taste when someone has a cold and the nasal cavities are clogged.

Visceral Sensations

Visceral refers to anything that involves the body's internal organs, such as the glands and the smooth and cardiac muscle. **Visceral sensations** are the result of internal changes. Two important visceral sensations are hunger and thirst. The receptors that detect hunger and thirst are believed to be specialized cells located in the brain's hypothalamus. Hunger receptors function by detecting deficiencies in blood nutrient levels, and thirst receptors look for deficiencies in the body's water content, or the body fluid's water-salt proportion.

People are not conscious of the hypothalamus detecting hunger and thirst because the brain projects these sensations. Thirst is projected to the mouth and throat, which will feel dry because less saliva is produced. Hunger is projected to the stomach, which contracts and feels empty. Usually people satisfy both sensations by eating and drinking. However, if hunger is not addressed with food, eventually the brain adapts and the hunger gradually decreases in intensity. This is because even though blood nutrient levels will decrease and prompt the hunger sensation, these levels eventually stabilize as fat from certain body tissues is converted to energy. Once this stabilization occurs, there are few changes for the receptors to detect, and hunger diminishes. However, the brain does not adapt if the thirst sensation is ignored. The body has no ability to stabilize as the water content decreases. Without stabilization, changes and fluctuations continue, which receptors continue to detect. The thirst sensation increases, and dehydration may result.

Vision

Vision receptors are located in the eye, along with a refracting system that directs light rays to the vision receptors located in the retina.

The eyeball is protected by eyelids and lashes. Eyelids are able to open and close over the eye because they are made of skeletal muscle. Eyelashes border the eyelids and keep dust and other debris from the eyelids. In addition, there is a thin membrane called the **conjunctiva** that lines the interior of each eyelid. Many eye infections are forms of **conjunctivitis**, in which the conjunctiva becomes infected and inflamed, making the eyes red and itchy.

Located on the upper outer corner of the eyeball are the **lacrimal glands,** which produce tears that cleanse the eyes and keep them moist (Figure 8.22). Tears are taken to the eye's anterior region through small ducts.

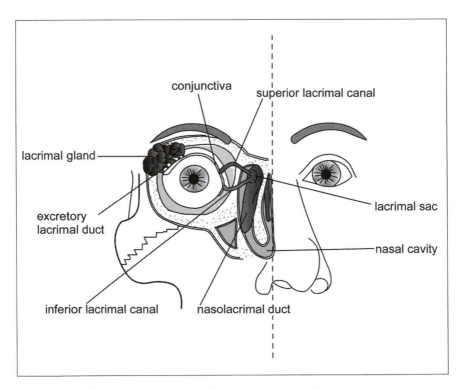

Figure 8.22 The Lacrimal Apparatus. The lacrimal apparatus is an important protection for the eye. (Sandy Windelspecht/Ricochet Productions)

Blinking spreads the tears and allows them to wash the eye. Composed mostly of water, tears also contain an enzyme called **lysozyme**, which prevents bacteria from producing on the eye's surface. In the outer portion of the middle of the eye are the superior and inferior **lacrimal canals**, which are ducts that transport tears to the **lacrimal sac**. Located in the lacrimal bone, the lacrimal sac leads to the **nasolacrimal duct**, which empties tears into the nasal cavity. This is what causes a runny nose when someone cries.

The **orbit** is a cavity in the skull that protects and surrounds the eyeball. There are six muscles that extend from the socket to the surface of the eyeball. These six muscles include four rectus muscles, which move the eyeball up and down and side to side. The remaining two oblique muscles allow the eye to rotate (Figure 8.23). These muscles function through the third, fourth, and sixth cranial nerves (oculomotor, trochlear, and abducens).

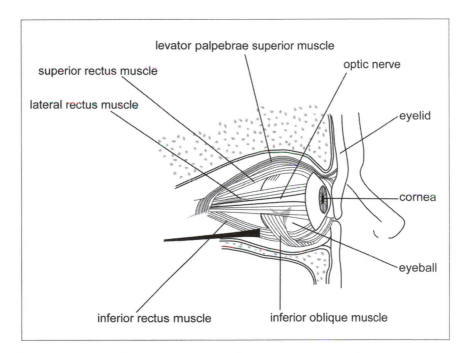

levator palpebrae superior muscle
superior rectus muscle
optic nerve
lateral rectus muscle
eyelid
cornea
eyeball
inferior rectus muscle
inferior oblique muscle

Figure 8.23 The Eye Muscle. This view of the eye displays vital muscles, in addition to the optic nerve. (Sandy Windelspecht/Ricochet Productions)

When examining the eyeball's anatomy, it is important to note the eyeball's three layers—the outer **sclera**, the middle **choroid** layer, and the inner **retina**. Composed of fibrous tissue known as the white of the eye, the sclera is the thickest layer. The **cornea** is located on the anterior of the sclera, and is unique from the rest of this layer because it is transparent and has no capillaries. This allows it to be the first part of the eye to bend (or refract) light rays.

The second layer of the eyeball, the choroid layer, is made up of blood vessels. In addition, this layer prevents glare by absorbing a certain amount of light within the eyeball. The outer portion of this layer contains the **iris** and the **ciliary body**, a circular muscle that is connected to the lens's edge by suspensory ligaments. Similar to the cornea, the lens is transparent and has no capillaries. The ciliary muscle allows the eye to focus light from objects near and far by changing the shape of the lens.

At the front of the lens is located the circular iris, which is known as the colored portion of the eye. The iris's opening is called the **pupil**. The pupil's diameter is controlled by two sets of muscle fibers. When the radial fibers contract, the pupil dilates or expands, which is a sympathetic response. The pupil constricts or reduces in size when the circular fibers contract, which is a parasympathetic response of the oculomotor nerves. This automatic or reflexive response is a protective mechanism because it prevents too much intense light from entering the retina. It also allows more precise near vision, which allows people to read books and other materials that are close to their eyes.

Another important part of the eye's anatomy is the retina, which is located on the interior of the choroid level (although it covers only two-thirds of the eye). The retina houses the visual receptors, called the **rods** and **cones** (Figure 8.24). Whereas rods only detect light, cones detect colors, which are actually made up of varying wavelengths of visible light. The **macula lutea** is abundant with cones and is located in the center of the retina behind the lens. The area known for the best color vision is the **fovea**, a small depression located in the macula lutea that contains only cones. Towards the edge of the retina is where the most rods reside. When light is dim, such as in a dark room or at night, we can best see through the periphery or sides of our visual fields, because this is where most of the rods are located.

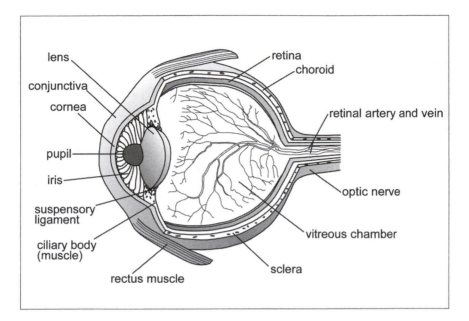

Figure 8.24   The anatomical elements of the eye. (Sandy Windelspecht/ Ricochet Productions)

Inside the eyeball are two cavities. Located between the lens and the retina, the larger posterior cavity contains a semisolid substance called **vitreous humor**. The retina is kept in place by the vitreous humor. However, if the eyeball is injured and the vitreous humor is lost, the retina can become detached. The second cavity, the anterior cavity, is between the front of the lens and the cornea. This cavity contains the **aqueous humor**, the eyeball's tissue fluid formed by capillaries in the ciliary body. Aqueous humor flows through the eye's pupil and is absorbed by small veins called the **canal of Schlemm**, located where the iris and cornea join together. Because the lens and cornea have no capillaries, they need the nourishment from the aqueous humor.

How the Eye Perceives Light

The vision process begins when the eye's receptors—the rods and cones— detect light and generate impulses. These impulses are then transported to the brain's cerebral cortex, where they are processed. An important part of the vision process is the **refraction** of light, which is when a light ray is bent or deflected as it passes from one object into another of greater or smaller density (Sidebar 8.3). In the eye, the refraction of light begins with the cornea, then the aqueous humor, the lens, and finally the vitreous humor. Adjustments can only be made in the lens and depend on the ciliary muscle. When the eye is trying to focus on a distant object, the ciliary muscle will be relaxed, making the lens thin out. But when trying to focus on closer objects, the ciliary muscle contracts, causing the lens to recoil and bulge in the middle, which gives the lens greater refractive abilities. Problems with the lens and **ciliary muscle** can be addressed with corrective lenses or glasses.

The next step in the vision process happens when light hits the retina, causing chemical reactions to occur in the rods and cones. The rods contain a chemical called **rhodopsin**. During a chemical reaction, rhodopsin breaks down into scotopsin and retinal, a derivative form of vitamin A. An electrical impulse is generated as a result of this chemical reaction. The chemical reactions in the cones also involve retinal and generate an electrical impulse. However, cones are also absorbing various wavelengths of light during this time. There are three types of cones: red absorbing, blue absorbing, and green absorbing. Every ray of light is taken in by one of these types of cones. Dysfunctions of the cones and rods can lead to night and color blindness.

The impulses from the rods and cones are carried by ganglion neurons to the optic disc, where they converge to become the optic nerve and exit the eyeball. Because the optic disc contains no rods or cones, it is sometimes called the eye's "blind spot." However, the eye is constantly moving and rotating to compensate for this blind spot. The optic nerves from the left and right eye join together at the **optic chiasma**, located right in front of the pituitary gland.

Fibers from each eye's optic nerve cross to the other side, allowing each side to capture visual impulses from both eyes. In the brain, the visual areas are located in the occipital lobes of the cerebral cortex. These visual areas integrate the slightly different picture transmitted by each eye into a single picture, which is called **binocular vision**. In addition, the image on the retina is actually upside down, but these visual areas correct this so people see the image right side up.

SIDEBAR 8.3

Correcting Vision Problems through Refractive Surgery

Vision problems are caused by errors of refraction. Normal vision is referred to as 20/20, which means that the eye can clearly see an object 20 feet away. If someone is nearsighted, or has myopia, the eyes can see near objects but not distant ones. For example, if someone has 20/80 vision, this means that the normal eye can see objects clearly at 80 feet away, but the nearsighted eye can only see that object if it is brought within 20 feet. Focusing of images by the nearsighted eye is done in front of the retina because the eyeball is too long or the lens is too thick. This can be corrected with glasses with concave lenses that spread out the light before it hits the eye.

When an eye sees distant objects well, it is farsighted, or hyperopic. For instance, vision might be 20/10, which means that it can see at 20 feet what a normal eye can see at 10 feet. This eye will focus images behind the retina due to a short eyeball or too thin lens. To correct hyperopia, glasses with convex lenses are used to unite light rays before they hit the eye. Contact lenses are also used to correct vision problems.

In addition to glasses and contact lenses, there are surgical procedures that improve the focusing power of the eye. These are a more permanent approach to vision correction. Here are some details on a few of these surgeries:

> LASIK (Laser-assisted In Situ Keratomileusis): Using a laser, this procedure permanently changes the shape of the cornea. The initial step is to cut a flap in the cornea using a blade or laser. This flap is folded back, exposing the midsection of the cornea known as the stroma. Then the laser is focused on the cornea, vaporizing part of the stroma. The flap is then replaced over the cornea.

> Radial Keratotomy (RK) or Photorefractive Keratectomy (PK): Both of these procedures also reshape the cornea. RK uses a knife to cut slits in the cornea. Like LASIK, PRK uses a laser (the procedure was developed prior to LASIK). While a flap is cut to expose the stromal layer in the LASIK procedure, the PRK exposes this layer

through the epithelium, which is the top layer of the cornea. The epithelium is actually scraped away to completely expose the stroma.

Thermokeratoplasty: This type of refractive surgery uses heat to reshape the cornea. While the source of heat can be a laser, the laser used is different from that used in LASIK and PRK. Another type of refractive surgery is in which heat is used to reshape the cornea.

Hearing

There are three main areas in the ear: the outer ear, the middle ear, and the inner ear (Figure 8.25). Receptors for hearing and **equilibrium**, or balance, are both found in the inner ear.

The **auricle** (or pinna) and **ear canal** make up the outer ear. Composed of skin-covered cartilage, the auricle is not important to humans, although it acts as a sound funnel for many animals, such as dogs and cats. Wearing glasses would be uncomfortable without an auricle, but it has no impact on hearing. The second part of the outer ear, the ear canal, acts as a tunnel into the temporal bone and middle ear. The **eardrum**, also called the tympanic membrane, stretches across the end of the ear canal and produces vibrations when hit with sound waves. These vibrations are transmitted to the three **auditory bones** called the malleus, incus, and stapes. This last bone, the stapes, transports vibration to the oval window through the inner ear.

Air enters and leaves the middle ear through the **eustachian tube**, also called the auditory tube, which extends from the middle ear to the nasopharynx. In order for the eardrum to function and vibrate, the air pressure in the middle ear must be equal to the pressure outside the ear. Ears often tend to "pop" when the air pressure is unequal, such as when in an airplane or when driving to a different elevation. The "popping" is caused when the eustachian tubes are trying to expand in order to equalize the pressure.

The inner ear is located within the temporal bone and encases a bone cavity known as the **bony labyrinth**. Lined with a membrane called the **membranous labyrinth**, this cavity contains fluid called **perilymph**, which is found between the temporal bone and the membrane.

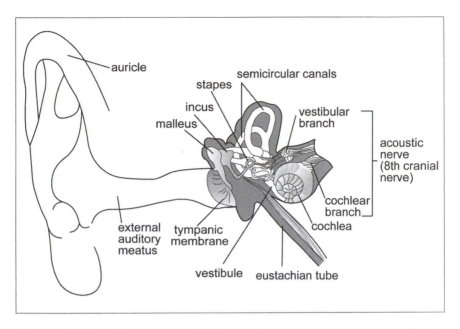

Figure 8.25 The inner, middle, and outer ear structures. (Sandy Windelspecht/Ricochet Productions)

Endolymph is the fluid found within the inner ear structures—the cochlea (important in hearing) and the **utricle, saccule,** and **semicircular canals** (all important in equilibrium).

The cochlea looks like a snail shell and is made up of two and a half turns that give it a coiled appearance. The cochlea is divided into three canals filled with fluid. The medial canal is known as the cochlear duct and encases the hearing receptors in the spiral organ, or **organ of Corti**. These receptors are known as hair cells, although they are not hair at all. These cells contain nerve endings from the cochlear branch of the eighth cranial nerve.

Hearing involves the reception of vibrations, the transmission of vibrations, and then the generation of nerve impulses (Figure 8.26). After sound waves enter the ear canal, they are transmitted to the ear structures according to the following sequence: eardrum, malleus, incus, stapes, the inner ear's oval window, the cochlea's perilymph and endolymph, and finally the organ of Corti's hair cells. Vibrations reach these hair cells,

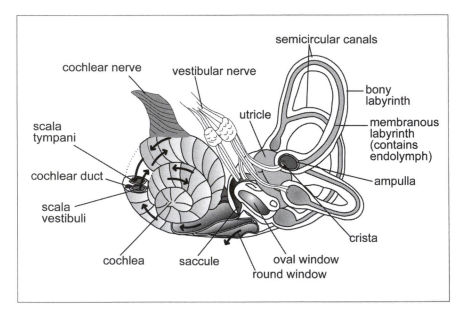

Figure 8.26   Inner Ear Structures. The inner ear structures are shown with details indicating the direction in which vibrations move during hearing. (Sandy Windelspecht/Ricochet Productions)

which bend and then generate impulses that travel to the brain through the eighth cranial nerve. Sounds are heard and processed in the auditory areas of the brain, which are located in the temporal lobes of the cerebral cortex. The hair cells are protected by the round window, which is found just below the oval window. The structure pushes out when the stapes pushes in through the oval window, thus relieving pressure and preventing damage to the hair cells. (Sidebar 8.4 contains an explanation of hearing loss.)

Two other inner ear structures, the utricle and saccule, are located in the vestibule between the cochlea and semicircular canals. These structures are actually hair cells surrounded in a gelatinous membrane with otoliths, which are tiny crystals of calcium carbonate. When the head changes position, gravity pulls down on these otoliths and bends the hair cells, thus generating impulses. The vestibular portion of the eighth cranial nerve then transmits these impulses to the cerebellum, midbrain, and temporal lobes of the cerebrum. At the subconscious level, the cerebellum and

---

SIDEBAR 8.4

Deafness

There are three different types of **deafness** (the inability to hear properly): conduction deafness, nerve deafness, and central deafness. Conduction deafness is when one of the ear's structures cannot transmit vibrations properly. This can be caused by a punctured eardrum, an auditory bone arthritis, or a middle ear infection in which an excess amount of fluid fills the middle ear cavity.

Nerve deafness occurs when there is damage to the eighth cranial nerve or the hearing receptors located in the cochlea. Some antibiotics can damage this cranial nerve. In addition, some viral infections, such as mumps or rubella, can also cause nerve damage. Nerve deafness often occurs in the elderly when hair cells in the cochlea become damaged from years and years of exposure to noise; this hair cell damage is accelerated by chronic exposure to loud noise.

Central deafness is when the auditory areas of the brain's temporal lobes become damaged. This can be caused by a brain tumor or other nervous system disorder.

---

midbrain interpret and process these impulses to maintain equilibrium. The cerebrum informs us of the head's position.

The last inner ear structure consists of three semicircular canals, which are also involved in stabilizing equilibrium. Each of these membranes is oriented in a separate plane and filled with fluid. At the bottom of each structure is the **ampulla**, an enlarged portion that contains hair cells sensitive to movement. When the body moves forward, the hair cells are initially bent backward and then straighten. Impulses are generated when these cells bend, and are also transmitted to the cerebellum, midbrain, and temporal lobes of the cerebrum via the vestibular branch of the eighth cranial nerve. The interpretation of these impulses is associated with stopping or starting, accelerating or decelerating, and changing directions. In general, the semicircular canals provide information while the body is in motion, and the utricle and saccule provide information while the body is at rest. The brain synthesizes all this information to create a unified sense of body position.

Receptors in the Bloodstream

Two of the heart's arteries, the **aorta** and **carotid**, have receptors to detect changes in the bloodstream. Blood pumped by the left ventricle is fed through the **aortic arch**, which works its way over the top of the heart. The **carotid arteries** are the left and right branches of the aortic arch that transport blood through the neck and then on to the brain (Figure 8.27).

Both of these blood vessel systems contain receptors. The **pressore-ceptors** are located in the carotid and aortic sinuses and detect changes in blood pressure; the **chemoreceptors** are located in the carotid and aortic bodies and detect changes in the oxygen and carbon dioxide content of blood. Rather than stimulate sensations, these receptors generate impulses that regulate breathing and circulation. For example, if there is a sudden decrease in the blood's oxygen content (known as **hypoxia**), this

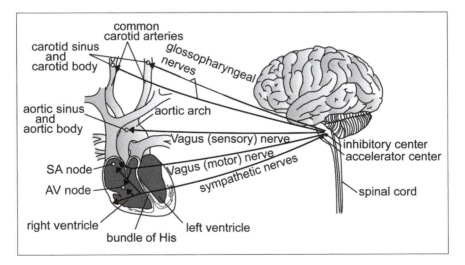

Figure 8.27   Regulation of the Heart by the Brain and Spinal Cord. The brain and spinal cord work to regulate various elements of the heart. (Sandy Windelspecht/Ricochet Productions)

change will be detected by the carotid and aortic chemoreceptors. These impulses are then transmitted through the ninth and 10th cranial nerves (glossopharyngeal and vagus) to the medulla. As a result of the sensory interpretation in the medulla, the respiratory rate and the heart rate will increase to collect and circulate more oxygen. This allows the body to maintain stable levels of oxygen and carbon dioxide in the bloodstream and maintain normal blood pressure.

Summary

The human body's nervous system has two divisions, the central nervous system (CNS) and the peripheral nervous system (PNS). The brain and the spinal cord make up the CNS, which is where the five senses are processed. The CNS detects a stimulus and facilitates the appropriate response in the effector organ or muscle. The PNS includes the cranial and spinal nerves, as well as the autonomic nervous system (ANS). The ANS has two divisions, the parasympathetic and the sympathetic divisions. Involuntary movements of the body's smooth muscle system are controlled by the ANS.

The foundations of the nervous system are the nerves and nerve fibers, which make up neurons. Neurons are the true activity hub of the nervous system—through the neurons, impulses and related messages are communicated throughout the entire nervous system.

9

The Reproductive System

Kathryn H. Hollen

Interesting Facts

- The proportions of a developing fetus change dramatically as it grows. The size of its head at the ninth week of development is almost one-half the total length of its body; an adult's, in comparison, is only about one-eighth the total length of its body.

- Increased age in women increases their likelihood of having twins because the zona pellucida tends to be harder in older women and causes the inner cell mass to break into clumps as it "hatches."

- About 50 percent of conceptions fail between fertilization and implantation due to abnormalities in the specialized cells required for implantation.

- A fetus can detect light coming from outside the mother's body; it will turn to follow a light moving outside and across the abdomen.

- The first successful vaginal hysterectomy for the cure of uterine prolapse was self-performed by a peasant woman in the seventeenth century. She slashed off the prolapse with a sharp knife, surviving the hemorrhage to live out the rest of her life.

- The brain of a child malnourished in the womb and in infancy may be 60 percent smaller than that of a normal child.

- Babies in the womb dream, which contributes to brain development.

- Several theories surround the discrepancy in size between the sperm and the egg. One is that the egg is larger because it contributes most of the bulk to the embryo while the small sperm must be an agile, mobile hunter of the ovum. Another suggests that sperm have to be tiny to be produced in quantity; the more produced, the greater the likelihood that only superior sperm will reach the egg.

Chapter Highlights

- Sex and reproduction

- Genetics

- Cell division

- Reproductive organs: female and male

- Endocrine system and sex differentiation

- Pregnancy: what happens in each trimester

- Prenatal Care: testing and pregnancy-related complications

- Preventing pregnancy

Words to Watch For

Alleles	Down syndrome	Histones
Asexual reproduction	Embryogenesis	Homologous
Blastocyst	Epiblast	Huntington's chorea
Chromatin	Gametes	Hypoblast
Circumcision	Gene expression	In vitro
Colostrum	Genetic imprinting	In vivo
Complementary base pairs	Genetic sex	Meiosis
Contraceptives	Glucose tolerance test	Mitosis
Daughter cells	Graft rejection	Monozygotic
		Morula

Multiple marker test

Oocytes

Oogonia

Phosphates

Polyspermy

Sex-linked inherited
   characteristics

Sickle cell anemia

Spermatogonia

Ultrasound scan

Ureters

Vestibule

Vestigial

Zona pellucida

Zygote

Introduction

The systems that animate the human body do not operate independently of one another. They are united by a network of blood, nerves, and tissues that give them substance and function, and, in this respect, the reproductive system is no different from its counterparts. But in another respect, it is unique, for it is the only system in the body dedicated exclusively to the continuation of the species. This single fact dictates a dual direction for the material in this chapter. On the one hand, it must address the biology of the reproductive system that keeps the human race going. On the other, it must also consider, albeit much more briefly, how tapping into human DNA can so transform genetic identity that the definition of what it means to be human may unravel.

Sex and Reproduction

Human beings have sex to reproduce, but there is another reason that is critical to the vitality of the species—and to that of other species as well. In fact, sex is a fundamental feature of the reproductive life of most animals, although some of them engage in it differently from humans. The frog, for example, lays her eggs near a river or stream for later fertilization by any compatible male frog that happens by, while humans customarily have reproductive sex together, at the same time. People are very different from frogs in another way, too, in that human sex can result in serious consequences such as unwanted pregnancy, disease, social disgrace, moral or religious disapproval, or legal difficulties. Asexual reproduction or cloning would likely be easier; hydra and sponges are able to grow buds genetically identical to themselves with far fewer consequences. Flatworms and sea stars reproduce by generating new organisms from their own detached parts. Why do people complicate their lives with sex and all that it entails?

The answer is simple: genetic diversity, the mingling of genes from two different parents to create genetically unique offspring. This is the very definition of sexual reproduction, a means of propagation that, unlike the asexual method, reduces the chance that harmful mutations will be passed on because the normal gene from one parent can dominate the defective gene from the other parent. Since not all mutations are bad, however, sexual reproduction is also an avenue for passing on beneficial traits, such as genetic adaptations that might, for example, improve one's resistance to environmental pathogens. And just in case the species neglects to engage in the reproductive activity that biology has so thoughtfully designed for it, renewed motivation arises from the persistent nudging of the libido—that compelling drive that, some would argue, is second only to the need for food and water in the urgency of its demand.

In a strictly biological sense, sex is fundamental to reproduction in humans, as in many other species, because the offspring gain genetic resistance to disease and threatening mutations are minimized or purged.

Genes

Genes, the human genome, genetic engineering, sequencing the genome— most people have heard and read a great deal about these subjects in recent years. Genes determine the characteristics that define the species as human: creatures who walk erect, who are born with two legs and arms and hands, and who are capable of learning to read and write.

Genes also determine if a person's legs will be short or long, his nose large or small, his hair blond or black; each of these traits makes every person unique, even though 99.999 percent of nearly all humans' genes are same. (Identical siblings like twins, however, start out in life with 100 percent of the same genes.) When deoxyribonucleic acid (DNA) replicates and accidentally drops a critical molecule from the gene during transcription, or when environmental assaults such as exposure to cigarette smoke accumulate in the cells' DNA, the damage can be serious enough that, later in life, the affected twin might develop a genetically related disease while the other does not.

Although it is generally understood that every person is the product of unique combinations of parental genes, not everyone understands the intricate details: what genes are made of, where they reside in the cell, and how cell division ensures genetic diversity in the species.

Like all living things, the human body is made up of chemicals, tiny molecules such as the four amino acids comprising genes. Called bases, these amino acids (adenine, cytosine, guanine, thymine) reside in the cell as part of the coiling DNA that forms diffuse nuclear matter known as chromatin. Two bases, adenine and thymine (A and T), always form one pair, and cytosine and guanine (C and G) always form the other. These are complementary base pairs that, held loosely together by hydrogen bonds, construct the "rungs" of DNA's familiar ladder or spiraling double helix. When a cell begins to divide, DNA twists and condenses out of chromatin to organize into chromosomes and to replicate. This is how the cell's copy will receive the same DNA—that is, the same chromosomes and genes. In turn, as these daughter cells divide, they produce exact replicas of themselves. Thus, all the genetic information of the first cell is transferred to every descendant.

Although the word "gene" is a singular noun, a gene actually represents the components of a given piece of DNA that act as a single unit to direct an activity. The genes utilize 20 different amino acids that, in turn, can construct about 250,000 different proteins. This is known as encoding; that is, the gene encodes for a protein, an enzyme perhaps, which in turn unleashes a cascade of chemical reactions triggering cellular activity. The gene that initiated this activity is said to have been expressed—that it has been "switched on."

Recent research indicates that proteins called histones embedded within DNA represent epigenetic information that tells genes when to turn on or off. In embryogenesis, the early stages of human development in the womb, they are switched on to tell the cells what kind of tissue to become—transforming a cell into a muscle cell, for instance, and instructing the new cell how to do its job. Every muscle cell division thereafter will produce a muscle cell, but it was a series of genes that told the first cell what to become (Figure 9.1). Cells in muscle tissue look different from those in epithelial tissue, just as both of these look very different from a neuron, a type of brain.

How the cells are instructed to become different kinds of cells and thus to take on different functions is a subject of intense interest, not just within the research community, but among the lay public, politicians,

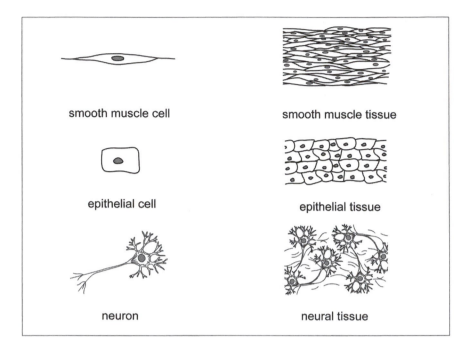

Figure 9.1   Cells and Tissues. Epithelial cells give rise to epithelial tissue, which forms the inside and outside surfaces of many organs; the muscle cell depicted here produces smooth muscle tissue, which comprises involuntary muscles like those of the intestines; and neurons (nerve cells) like these feature dendrites, branching arms reaching out from the body of the neuron to receive impulses from the nervous system. (Sandy Windelspecht/Ricochet Productions)

bioethicists, and the medical establishment. Because that first cell, called a stem cell, has the ability to differentiate into any other kind of cell the genes so order, its therapeutic promise is tantalizing. Researchers are developing techniques to cultivate the cells in the laboratory and to manipulate their differentiation with the goal of growing human organs—in vivo or in vitro—to replace diseased ones. But before therapeutic breakthroughs of this kind are likely to be commonplace, a significant ethical hurdle must be addressed: the stem cells with the greatest potential for differentiation are generally obtained from human embryos, although current research is looking into the exciting possibilities of "reverse programming" differentiated cells back into stem cells. Adult stem cells can be harvested for therapeutic purposes from certain portions of the body, but most scientists find them of limited value.

Sequencing the human genome simply means mapping all the genes—each specific group of bases that, once switched on, directs the manufacture of the proteins that drive cellular functions. While it is tempting to assign one function to one gene, it is usually not that simple, although there are instances, such as in **Huntington's chorea** or **sickle cell anemia**, in which one causative gene can be identified. Usually, however, it appears that different genes may interact with external features in the environment, or may combine to predispose someone to a disease, or, conversely, to confer resistance to it. Nevertheless, links can be made between the influence of specific genes on specific characteristics or diseases, and some of these will be discussed later in this chapter.

Interestingly, although the human genome contains roughly three billion base pairs, scientists disagree about how many of these are genes. Researcher believes that there is a great deal of excess DNA, called introns or "junk DNA," whose role puzzles scientists. Many believe that these non-coding sequences, probably enough in each person's 100 trillion cells to reach the sun and back, are made up of meaningless pieces of viruses and other genetic debris whose sole value lies in providing an evolutionary fossil record of human ancestry. Others maintain that they may represent crucial coding sequences that control other genes in as-yet-undetermined ways and thus multiply the effect of all.

Chromosomes

The DNA "ladder" has two spiraling "uprights" made up of phosphates, sugars, and bases. One phosphate, sugar, and base form a nucleotide that matches up to its complement, another group of three molecules; if one piece of the DNA is made of phosphate, sugar, and adenine (one of the four bases), its counterpart would be attached at the rung to thymine, sugar, and phosphate. In every cell (except mature red blood cells, which discard their nuclei during human development to carry more oxygen throughout the blood), about six feet (180 centimeters) of DNA resolve out of chromatin into 46 coiled chromosomes that form 23 pairs. Each pair is comprised of two homologous chromosomes—that is, they carry essentially the same genes, but one chromosome is from the mother and one chromosome is from the father. An example is the gene for eye color. Suppose a person inherited a gene for blue eyes from her mother and a gene for brown eyes from her father. The genes for eye color in this case are called alleles because, while they are essentially the same gene—they both transmit eye color characteristics—one encodes for one color on the mother's chromosome and the other for a different color on the father's. Thus alleles are different expressions of the same gene on homologous chromosomes.

So, 23 chromosomes from one parent pair up with their homologs from the other parent. Each chromosome pair, however, is different from every other pair. Twenty-two of these pairs are called autosomes; the 23rd pair comprises the sex chromosomes, so called because they determine the sex of the offspring and carry the genes for sex-linked inherited characteristics like color-blindness. In women, the sex chromosome is called an X chromosome; in men, it is a Y chromosome.

Cell Division: Meiosis and Mitosis

Humans have two kinds of cells: body (somatic) cells, which form all of their tissues, and sex (germ) cells, which in women are eggs and in men are sperm. Although a human being's body cells are building and repairing tissues like heart muscle or bones throughout that person's life, sex cells have prepared themselves for one thing only—reproduction. The decision about which cells became body cells and which became sex cells was a differentiation decision made early in embryogenesis. While it is true that all but the red blood cells contain exactly the same 46 chromosomes, there comes a time when the sex cells must decrease their number to 23, or a haploid (n) number, so that, once male and female sex cells unite in fertilization, the original diploid number of chromosomes, 46 (or 2n), will be restored in the new organism.

Sex cells divest themselves of half of their chromosomes, and chromosomes mix up their genes to impart diversity to new cells, during a special type of cell division called meiosis, or meiotic division. Like regular cell division, meiosis relies partially on ribonucleic acid (RNA), a chemical closely related to DNA that might be, some suggest, DNA's evolutionary precursor. In humans, RNA carries out essential missions such as relaying messages from the genes to the cells and replicating DNA so there is a copy available for the new cell.

The function of meiosis is to mix both parents' genes and to reduce the number of chromosomes in the resulting cells by half; whereas, in mitosis (mitotic division), the full complement of chromosomes from the parent cell is passed on to the daughter cell with no change in genetic material. Thus, meiosis ensures the progeny will receive a diverse set of genes and, at the end of meiotic division, the germ cells will have become gametes, reproductive cells containing only 23 chromosomes. When meiosis is completed, the zygote will inherit traits from both mother and father because the nucleus of its father's sperm contained 23 chromosomes that were comprised of genes the father received from his mother and father, and the nucleus of its mother's egg also contained, at fertilization, 23 chromosomes that carry genes she received from her mother and father. In this way, the zygote received 46 chromosomes' worth of its parents' (and their parents', and so on) genetic material. Someday, as a sexually mature adult, what is now the zygote may contribute 23 of its chromosomes to its own child.

In the male, there is always a sufficient supply of sperm maturing in the testes, because meiotic division has already been completed and four haploid sperm cells from each germ cell reside there. In the female, however, a specific number of egg cells are produced only by mitotic division early in the development of the female embryo, and it is not until the maturing female reaches puberty that, at ovulation, one egg a month will complete its first meiotic division and discharge excess chromosomes not into a second egg, but into a polar body, a poor cousin to the ovum. Meiosis is arrested at that point, and not until fertilization will the ovum complete its second meiotic division, retaining most of the original cellular cytoplasm for itself while the extra chromosomes and remaining cellular material are consigned to another polar body or two, all of which are nonfunctional and will simply degenerate. Immediately after fertilization, the pronuclei of the sperm and egg cells merge and become the nucleus—now with 46 chromosomes—of the newly conceived organism, the zygote.

The Reproductive Organs

The last section briefly discussed differentiation—how certain cells become nerve cells, others become muscle cells, and still others become sex cells. In the early embryo, the latter are called primordial germ cells. They migrate to an undifferentiated area of tissue called the genital ridge where they multiply, becoming oogonia in the female embryo's developing ovaries or spermatogonia in the male's developing testes, or testicles. The essence of human life, they are poised to complete their development whenever the body summons them.

Female Reproductive Organs

Although several million primordial germ cells are produced during the embryo's development in the womb, hundreds of thousands die en route to the ovary and afterwards, during a girl's growth. This is due to apoptosis, or cell suicide, the body's normal biological response to excessive cell proliferation. A female born with about 2 million eggs in her ovaries is left at puberty with 300,000 to 500,000. Of these, she will use only 400 to 500 throughout her reproductive life. Until recently, scientists believed that women cease egg production forever at menopause. Intriguing new findings in mice, however, suggest that ovarian stem cells continue to produce eggs throughout the animals' lives. If the same is true in humans, it could have profound implications for infertility and aging in women.

The ovaries housing the eggs are two almond-shaped organs, one on either side of the pelvis in the lower abdomen, attached to the uterus by ligaments (Figure 9.2). At any one time, oocytes may be maturing in the ovaries in preparation for eventual fertilization. Regulated by hormones, follicles form around each of the oocytes, one of which matures into a Graafian follicle containing what is by then a secondary oocyte.

At ovulation, the mature ovum bursts from the ovary and the next follicle begins to mature (Figure 9.3). Once the egg is released, it is drawn into the Fallopian tube, or oviduct, that leads into the uterus. Each oviduct, about 5 inches (12.7 centimeters) long, has fringelike projections called fimbria that help direct the egg. New studies suggest that the egg's location in the Fallopian tube is a slightly warmer area than surrounding tissue, and that sperm have heat sensors to guide them to the site. If the

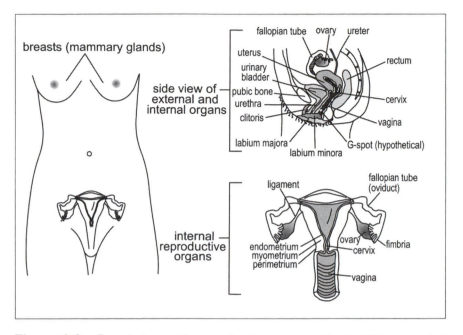

Figure 9.2   Female internal reproductive organs. (Sandy Windelspecht/ Ricochet Productions)

egg is fertilized, it completes the meiotic division that was arrested at ovulation. Fine hairs called cilia coordinate with the contractions of the oviduct to propel the egg down to the uterus, or womb, the upside-down pear-shaped organ that will house the developing fetus.

If the egg is not fertilized, the uterus sheds its endometrium, a special lining that thickens in preparation for the fertilized ovum, and expels it during menstruation along with the egg. Powerfully muscular, the uterus is capable of expanding to the size of a basketball during pregnancy. Its wall has three principal layers: the endometrium; the myometrium, the complex of muscles that surrounds the uterus to contract in childbirth and to help reduce uterine size afterwards; and the perimetrium, the outer layer of connective tissue.

The lower third of the uterus narrows into the cervix and the vagina. An organ about one inch in circumference, the cervix protrudes into the upper cavity of the vagina and dilates during childbirth to permit the infant's head to emerge. Its opening into the vagina, the external os, is

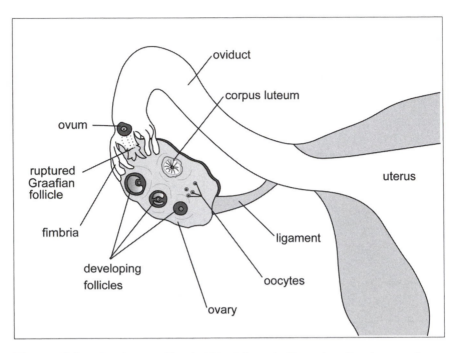

Figure 9.3   Ovulation. (Sandy Windelspecht/Ricochet Productions)

covered with cervical mucus that changes with the cycling of the menstrual cycle. During pregnancy, it thickens, forming a plug to keep out threatening pathogens. Also known as the birth canal, the vagina is a muscular cavity 3 to 5 inches (7.6 to 12.7 centimeters) long that is ordinarily narrow and collapsed on itself, but it expands and lubricates to admit the erect penis during intercourse. If pregnancy occurs, it stretches much more to accommodate the fetus during childbirth.

While it is not technically part of the genitalia, the urethra is so intimately associated with both the female and the male reproductive tracts that it merits a brief description here. In the female, urine produced by the kidneys flows down the ureters into the bladder, which stores and then releases it through the urethra to exit the body at the vestibule.

There are hidden but important glands on either side of the vagina called the lesser and greater vestibular glands (Bartholin's glands) that secrete fluids into the labia to help lubricate them during sexual arousal. There are two pairs of labia; the labia majora, or larger lips, are the fleshy outer lips that cover the more sensitive, hairless labia minora, or smaller lips.

Just behind the urethra, tucked within the labia, is the vaginal opening, which in young girls is partially covered by a thin membrane called the hymen. Although this tissue is usually present in those who have never had intercourse, some girls are born without it or unknowingly tear it during physical activity or with the repeated use of tampons. The small expanse of skin between the vaginal opening and anus is called the perineum. Each female is born with an organ devoted entirely to sexual pleasure, the clitoris. Comparable to the penis in terms of sensation, the exquisitely sensitive clitoris is hooded by a prepuce (a membranous, protective tissue capable of sliding back, or retracting). The glans is the tip of the clitoris that can be seen at the upper junction of the labia minora. The rest of its body is the shaft, which disappears into the pelvis and consists of two cavities that fill with blood during sexual excitement. Some insist there is a knot of tissue within the vagina known as the Gräfenberg spot or "G spot," a controversial subject because recent evidence suggests it does not exist. But others believe it to be a sensitive site of female pleasure located an inch or so into the vagina on the front wall, adjacent to the urethra and bladder and behind the pubic bone. The external female genitals are referred to collectively as either the vulva or the pudendum. Both terms include the mons veneris or mons pubis, the mound of tissue lying over the pubic bone that in sexually mature women is covered with pubic hair.

The breasts are mammary glands that, despite their undeniable sexual significance, evolved primarily to feed the young. The nipples' so-called erectile tissue—although susceptible to stimulation—is altogether different from that of the penis or clitoris. Smooth muscles within the areola, the differently colored tissue surrounding the nipples, are responsible for the erectility, which probably helps infants find and grasp the nipples more easily.

When she enters puberty, a young girl's breasts begin to accumulate fat and grow. A radiating series of lobes in each breast that reduces to smaller lobules and ends in tiny sacs called alveoli drain into lactiferous ducts carrying milk to the nipples. During pregnancy, high levels of estrogen and progesterone cause the alveoli to fill with proteins and other fluids. When the infant is born, the breasts release these nutrients, or colostrum, which impart the mother's immunities to the newborn. Within just a few more days, the breasts begin to lactate (produce milk). Nursing not only feeds the child; it also triggers powerful contractions in the mother's uterus that help it revert to its normal size.

When ovulation occurs, the Graafian follicle from which the secondary oocyte erupts is transformed into a corpus luteum, a body that produces the hormones estrogen and progesterone to prepare the uterus for implantation of the fertilized egg. These hormones, under the direction of the pituitary gland, orchestrate the phases of a woman's monthly menstrual cycle, 28 days long on average. At mid-cycle, ovulation, the lining of the uterus thickens and new blood vessels grow. Ten to fourteen days afterwards, if the egg is not fertilized, the corpus luteum ceases hormone production and degenerates; in response, the endometrium breaks down and is expelled from the body in the form of menstrual blood and tissue. The monthly menstrual cycle begins anew with the maturation of another primary oocyte in one of the ovaries. Although nature's choice of which ovary is released from the ovum each month is entirely random, evidence indicates an equal distribution of labor—each ovary seems to contribute about half of the total eggs ovulated during a woman's reproductive life.

Male Reproductive Organs

Spermatogonia, or immature sperm, also undergo changes before they are capable of fertilizing an egg (Figure 9.4). They reside in some 800 feet (24,000 centimeters) of seminiferous tubules, subsequently dividing by meiosis to become spermatids. Then, under the direction of specialized Sertoli cells, they differentiate and mature into spermatozoa, or sperm, several hundred million of which will be made in the testes daily from puberty onward. Since sperm must be kept at the right temperature to remain alive and healthy, the testes are nestled within a pouch of skin, the scrotum, suspended outside the body. In cold weather, the scrotum contracts to pull the testicles closer to the body for warmth. Too much heat can also damage sperm; tight clothing that restricts air circulation or regular bathing in water that is too hot can kill enough sperm to imperil fertilization.

At ejaculation, mature sperm stored in testicular ducts called the epididymides enter the vasa deferentia (singular form: vas deferens) that extend up into the body and behind the bladder. There they meet the seminal vesicles,

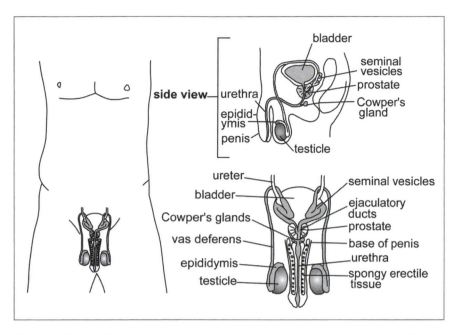

Figure 9.4 Male external and internal reproductive organs. (Sandy Windelspecht/Ricochet Productions)

glandular structures that narrow with the vasa deferentia into ureters, and form ejaculatory ducts. Sitting just below the bladder and surrounding the ureters is the prostate, a gland that contracts during ejaculation and secretes fluids into the urethra. This fluid combines with sperm and the secretions of the seminal vesicles and the bulbourethral, or Cowper's, glands to make semen, the whitish fluid ejaculated through the urethra at the end of the penis.

Like the clitoris, the penis contains spongy tissue whose vessels become engorged with blood during sexual arousal; this causes an erection that allows the penis to penetrate the vagina and deposit semen during ejaculation. Semen protects sperm in several ways: it provides a safe and fluid environment, it helps neutralize harmful acids in the male urethra and female vagina, and it supplies sperm with needed energy to swim up and into the uterus, a journey that must be completed within 12 to 72 hours before they die. Although millions of sperm are contained in a teaspoon or two of semen (the average amount in each ejaculate), fewer than 1,000 will reach the Fallopian tubes.

Circumcision

Males and females alike are born with a prepuce that covers the glans of the penis and clitoris, respectively. In males, the prepuce is also called the foreskin. It is attached to the glans but usually, by the time boys are of school age, the foreskin has naturally separated from the penis except for one anchoring point at the frenulum from which it can retract. Many males undergo a procedure to remove the foreskin surgically at birth; this is known as circumcision. Within certain societies around the world, girls and women are circumcised too, in what is usually an excruciatingly painful and mutilating process carried out to allow males lifetime power over female sexuality.

Male Circumcision

Male circumcision was first performed centuries before the beginning of the common era (BCE), and many cultures since have continued the practice for hygienic, cultural, or religious reasons. Although the procedure is performed nowadays in a clean environment by a trained person, there is a great deal of controversy surrounding it. Because circumcision usually takes place with no anesthesia when one is an infant and has no say in the matter, and because the absence of a foreskin may deprive men of significant sexual pleasure, many regard it as an unnecessarily cruel procedure akin to mutilation. On the other hand, there is evidence that uncircumcised men are more vulnerable to urinary tract infections and sexually transmitted diseases (STDs) than their circumcised counterparts, and they bear a slightly increased risk of penile cancer as well.

The Nervous System

The Brain and Spinal Cord

The brain is the most important sexual organ in the human body. Partnered with the spinal cord to make up the central nervous system, it not only supervises the nervous and the hormonal, or endocrine, systems that regulate the physiology of sexual response (what happens chemically and physically in the body), it is also the repository of the images, thoughts, and feelings that humans associate with sex.

The oldest neural tissue in the human brain, the reptilian brain is mediated by the influence of mankind's thinking brain, or neocortex. A third area, the limbic brain, may be thought of as a bridge between the two, and the place where emotions and sexual impulses reign. The brain is much more complicated than this; brain structures are coordinated in unaccountably complex ways to balance human instincts and emotions with appropriate behavior. There is nevertheless compelling evidence that some aspects of sexuality are "wired" into the brain. Evolution has etched into humans' unconscious many of the characteristics that are desirable in a mate, and these traits subtly influence how they choose one another. Men may look for the "right" proportions in women, a certain waist-to-hip ratio that suggests a suitable physique for pregnancy and sufficient fat deposits to nourish the young. Women, on the other hand, may tend to seek men who are tall, perhaps because their increased height once conferred hunting advantages and thus meant they would be better providers for the family. In most cases, men and women are not consciously aware of making these choices; they know only that they find someone sexually attractive.

The Peripheral Nervous System and Neurotransmitters

The first arm of the nervous system, the brain and spinal cord combination, works in concert with a second arm, the peripheral nervous system, made up of neural networks that thread throughout the body. It is divided into two parts, the somatic and the autonomic. The somatic sends sensory signals to the brain and activates certain kinds of motor activity, but in human reproduction, the autonomic nervous system (ANS) is the main player. It is always working, orchestrating involuntary functions outside of one's control. Very simply put, its sympathetic system makes preparations for the body to go into action, and its parasympathetic system reverses the preparations. (A third part of the ANS is the enteric system, which influences certain functions such as digestion.) When sexual arousal occurs, the sympathetic system increases pulse rate and saliva production, quickens breathing, and raises blood pressure; as excitement ebbs, the parasympathetic system slows pulse rate and breathing, lowers blood pressure, and decreases glandular secretions.

But this is by no means the whole story. Communication among the cells of the body is an extraordinarily complex process that relies on chemical messengers. Some of these are neurotransmitters, made by nerve cells to carry messages in the form of impulses across the synapses, or gaps, that separate them. They convey emotions, thoughts, ideas—all the neural processing that animates humans—and help regulate the secretion of hormones. Some are called the "feel-good" neurotransmitters: dopamine, acetylcholine, and the endorphins, known as natural painkillers. These, along with other biochemicals like norepinephrine and serotonin that act as neurotransmitters in the brain, seem to have major roles in sexuality and reproductive functions, probably due to their regulatory effects on mood. An imbalance or deficiency in any one of these neurotransmitters can result in depression, lethargy, insomnia, anxiety, or difficulty with concentration. The body carefully governs this delicate mix primarily through the hypothalamus. The hypothalamus is a regulatory organ in the brain that maintains conditions in the body like temperature, metabolism, and blood pressure, and oversees the limbic system, where emotions such as aggression and rage reside. It also issues critical instructions to the pituitary, the master gland that, directly or indirectly, dispenses the body's hormones.

Hormones

Hormones are another kind of biochemical messenger. There are two types, steroid and nonsteroid, that are secreted by glands throughout the body that comprise the endocrine and exocrine systems (Figures 9.5 and 9.6). By means of a biofeedback mechanism, the hypothalamus is alerted to abnormal biochemical levels; in response, it secretes neurohormones to tell the anterior and posterior lobes of the pituitary to increase or decrease the relevant hormones (Sidebar 9.1). For example, when the hypothalamus releases the neurohormone gonadotropin-releasing hormone (GnRH) to the anterior pituitary, it is telling the gland to produce both follicle-stimulating hormone (FSH) and luteinizing hormone (LH), chemicals that the ovaries and the testes require to support egg and sperm development. Once the pituitary delivers the message, the ovaries and testes begin to do the work to which they have been hormonally assigned. An example of a hormone produced by the posterior pituitary is oxytocin, related to uterine contractions, milk production, and emotional bonding.

Steroid hormones are made by the adrenal glands in both males and females. These chemicals are converted primarily to estrogen in women and, to a lesser degree, to testosterone; in men, they are converted primarily to testosterone and, to a lesser degree, to estrogen. These, along with

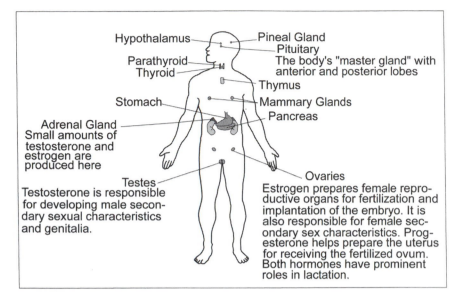

Figure 9.5   Glands of the human body. (Sandy Windelspecht/Ricochet Productions)

progesterone, are called the sex hormones because they are also produced in the ovaries and testes and are fundamental to reproductive biology.

The Role of Hormones in Embryonic Sex Differentiation

When fertilization occurs, the chromosomal arrangement of the merged gametes determines the genetic sex of the child. An X ovum fusing with a Y sperm yields an XY embryo, or a male. An X ovum fusing with an X sperm yields a girl. But the "default" embryo—the newly conceived embryo that has not yet developed sexual organs or produced hormones—is generally considered female until other steps involved in sex differentiation take place. It is critical that each step occurs at the right time, or normal development will go awry.

A seven-week-old XY embryo already has testicular tissue identifying him, at the level of his gonads, as male, just as an XX embryo has ovarian tissue identifying her as female. Both embryos have two ducts, the Wolffian and the Müllerian, that will be transformed into the reproductive structures

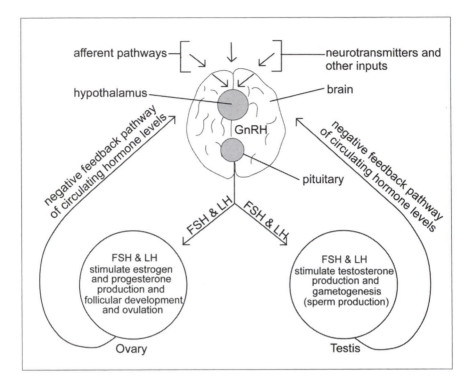

Figure 9.6   Biomechanical pathways. (Sandy Windelspecht/Ricochet Productions)

of either the male or the female, respectively. Scientists believe that a gene expressed only in the embryonic male brain triggers the development of testes; the early testicular tissue secretes testosterone, the transforming agent that causes the Müllerian duct to degenerate and prompts the male genito-urinary tract to develop. In contrast, without that testosterone acting on the XX embryo, the Wolffian duct degenerates, leaving the Müllerian to give rise to the female's reproductive organs.

Just as hormones are associated with the differentiation of the gonads and external reproductive organs, so are they associated with sex differentiation in the brain. Once again, testosterone plays an especially critical role.

Remember that the brain of the embryo is still that of the "default" gender, even though male sexual organs may have begun development.

SIDEBAR 9.1

The Hypothalamus and Mechanisms of Feedback

Also known as the body's "guardian," the hypothalamus ensures homeostasis, or equilibrium, within the body. It provides a link between the nervous and endocrine systems to regulate temperature; maintain appropriate hormone levels; signal sensations of hunger, fullness, and thirst; juggle aggression, fear, and rage; and set circadian (sleeping versus waking) rhythms. In reproduction, it triggers the pituitary to prompt sex hormone production by the ovaries and testes.

While it is not completely understood how the hypothalamus receives all of its input, it is known that several afferent pathways carry neurotransmitters and biochemicals into the organ from the brain and body. To help regulate reproductive functions, it relies on a feedback system; negative feedback is a function of hormone levels. For example, reduced blood concentrations of estrogen and progesterone that result from the degeneration of the corpus luteum after ovulation alert the hypothalamus to a low level of these essential hormones. The hypothalamus releases GnRH that instructs the pituitary to secrete FSH and LH to stimulate ovarian production. As blood levels of the newly produced hormones then rise, the hypothalamus stops triggering their secretion. This mechanism is illustrated in Figure 9.6.

To masculinize the brain, a certain amount of estrogen (testosterone that has been converted in the male's body) must pass through the brain barrier. It is not entirely clear why it is the primarily "female" hormone estrogen that is required, but one thing is very clear—estrogen must not be permitted to penetrate the female's brain barrier at this stage. A protein the female embryo produces, called alpha-fetoprotein, erects a barrier of its own that successfully blocks the estrogen's entry.

This process does not mean that estrogen is responsible for every "male" thought a man possesses. While it seems likely that hormones regulating sex differentiation are responsible for some of the structural differences between the male and female brain, it is also clear that social or cultural factors affect its fundamental development as well. During their youth, boys and girls are bombarded with input that will shape their individual sexual identities. General attitudes as well as individual experiences can influence them; today's "ideal" woman is thin, for instance, but before World War I, with a few exceptions, the plumper, rounded female figure was widely favored. However, there is consensus that even if a given input has no intrinsic sexual content, it could be integrated differently by a male brain than by a female brain, simply because their neurological networks are different. Analyzing what this means in terms of human behavior is not the focus of this chapter, but the implications offer a tantalizing glimpse into how profoundly the brain affects human sexuality and how differently it shapes male and female perceptions.

The Libido and the Physiology of Sexual Response

The sex drive, or libido, is one of the most compelling drives in human experience, in part because it ensures continuation of the species. It is a source of great pleasure and fulfillment, but it can also be responsible for great heartbreak and loss—even, in extreme cases, violence. People under its influence have jeopardized marriages, parental rights, job security, economic status, national or international prestige, and countless other symbols of stability and respectability for what is sometimes only a "one-night fling."

The first phase, excitement, is the period when sexual stimulation activates the sympathetic nervous system to increase heart rate, produce hormonal secretions, and develop tissue swelling and erection. When someone responds to stimuli, the sympathetic arm of the ANS instructs the Cowper's glands in the male to secrete fluid; it tells the arterioles, the blood vessels that carry blood from the heart, to dilate, opening them wider to allow greater blood flow. At the same time, it orders the venules, small veins carrying blood back to the heart, to constrict and thus prevent too much blood from returning; in this way, blood accumulates in the genitals and, in the male, engorges spongy tissue in his penis to cause an erection. Similarly, in the female, engorgement causes her clitoris to become erect while her vestibular glands secrete lubricating fluid and her pulse rate and blood pressure rise. In addition, her breasts and the outer portion of her vagina may swell as her labia become darker in color.

The second phase, plateau, is an intensification of excitement; vascular congestion in the genitals is at its peak, flushing of the skin might occur, and muscles of the thighs and buttocks tighten. Interestingly, the clitoris retracts under its prepuce in this phase and shortens by up to 50 percent, a phenomenon that seems to signal the onset of female orgasm.

The third phase is orgasm, or climax. The male reaches ejaculatory inevitability, or "the point of no return" at which he can no longer delay powerful spasms that originate in the epididymides and pulse through the vasa deferentia and prostate. The entire nervous system becomes involved; this is the height of his pleasure, and the moment that sexual tension is discharged amid strong contractions that ordinarily lead to ejaculation. But not always. Men may also have a "dry orgasm," an orgasm with no expulsion of seminal fluid. This syndrome can accompany certain neurological conditions or diabetes, it can occur in perfectly healthy males who have ejaculated frequently over a recent period of time, and, most commonly, it happens to young boys nearing puberty when they are experiencing sexual pleasure but their semen-producing organs have not fully matured. Retrograde ejaculation, when semen backs up into the bladder, is another reason for dry orgasm. In the absence of disease, this is a normal phenomenon following several episodes of arousal that do not result in orgasm; fluids quite literally back up into the bladder and subsequently must be released during urination.

Physiologically, the female's orgasm is similar to the male's in terms of buildup and the point at which contractions begin. These are overwhelmingly pleasurable, pulsing through her uterus and vagina in wavelike patterns. Although her orgasm is not necessary for fertilization, uterine spasms in the outer third of the vagina can dip the cervix into contact with sperm and may even help draw sperm up into the cervix, increasing the chance of fertilization. Because she produces no sperm, however, her pleasure and orgasm are almost incidental to the biology of human reproduction; if her menstrual cycle deems her ready for fertilization, there is every possibility she will be impregnated even as a result of rape, no matter how brutal her experience.

Some schools of thought hold that there are two types of female orgasm, vaginal and clitoral, the latter, according to Freudian psychoanalytic theory, being the more "immature," and that an individual woman is capable of experiencing only one kind. Others maintain that the same woman, depending in part on how and where she is stimulated, can experience both. Some women report that an orgasm felt deeply in the vagina is the more intense of the two, triggering stronger uterine and vaginal contractions, while other women say that sensations arising from clitoral stimulation are more pleasurable. All must agree, however, that it is actually the brain that reigns supreme in the sexual response department, for only the brain can trigger a spontaneous orgasm. A sleeping person receiving none of the physical or external stimulation on which arousal normally relies can have an orgasm simply because his or her brain created some exciting imagery. In fact, so complete is the brain's mastery over arousal that some women can fantasize to orgasm while fully conscious, using nothing for stimulation but the rich material her neural activity evokes.

The final period of a sexual encounter is called resolution, when the parasympathetic system begins to reverse the excitement that its counterpart ignited and allows blood from engorged genitals to ebb. Norepinephrine, one of the feel-good hormones, is released, adding to the overall sense of well-being and relaxation that the participants enjoy.

From Sex to Pregnancy—or Not

Despite scientists' sophisticated understanding of human reproductive biology, conception, the beginning of life, cannot be defined. Just because a sperm has penetrated an ovum does not mean fertilization has taken place; that occurs only when the pronuclei of the sperm and egg (two gametes, each with 23 chromosomes) fuse to create a zygote, whose nucleus will then contain 46 chromosomes.

Some view fertilization as the moment of conception. Others feel that fertilization merely launches cell division, producing a nondescript cluster of cells that is not alive until it implants in the womb, establishes a nourishing blood supply, and begins to differentiate into human tissue; these events, they believe, represent conception. So to avoid any misunderstanding, fertilization, rather than conception, is used here to describe the beginning stages of embryogenesis.

Most biological processes build on one another and balance complex cellular and hormonal activity with split-second timing. Nowhere is this more evident than in a human being's first three months in the womb.

Developments during the last six months are much less dramatic but are equally important, because they represent organ maturation and growth necessary for life outside the mother's body.

Every parent hopes for a normal pregnancy that ends with the birth of a healthy child, and the following describes how such a pregnancy might unfold in the context of a hospital or home delivery attended by physicians or midwives. In addition, prenatal care, fetal and maternal screening, and frequent medical issues pregnant women encounter are discussed.

Many women choose not to have children. In recent decades, thanks to dependable contraceptives and safe, legal abortion, a decision to postpone or avoid parenthood entirely can easily be made. This section concludes with a description of the methods used to prevent or terminate unwanted pregnancies.

The First Trimester

Fertilization

Given the long and dangerous journey each sperm makes, from the upper vault of the vagina through the cervix and uterus and into the Fallopian tubes where it encounters the ovum, it is surprising that fertilization occurs at all. But out of the millions of sperm ejaculated during intercourse, a few hundred or so do indeed survive the journey and, in as little as 15 minutes or as much as 72 hours, reach the upper Fallopian tubes where the ovum awaits, abundantly covered with sperm receptors. Normally, only one sperm can penetrate the ovum; the moment its surface enzymes digest a path through the ovum's outer layer, the zona pellucida, the ovum dispenses enzymes of its own that break down its receptors and harden its outer layer to make it impenetrable. In a rare event known as polyspermy, more than one sperm breaks through. Because this leads to altered chromosome numbers that would result in abnormal development, the ovum fails to develop and is expelled (see Sidebar 9.2).

Although fertilization occurs just after ovulation, the medical community usually calculates pregnancy, or gestation, to begin at the start of a woman's most recent period and to last 280 days (rather than the 266 days elapsing between fertilization and childbirth). The 280 days or 40 weeks are divided into three trimesters, the first spanning the weeks between

SIDEBAR 9.2

The Question of Twins

Polyspermy raises the question of twins: if only one sperm can fertilize an egg, how do twins develop? The twofold answer lies in the difference between fraternal and identical twins.

Fraternal Twins

These children are not identical and may be male and female; they result from two different eggs being fertilized by two different sperm. In a given month, a female may ovulate from both ovaries; each of her two Fallopian tubes contains a mature ovum awaiting fertilization. After intercourse, a different sperm fertilizes each. So two genetically distinct siblings—brother-brother, sister-sister, or brother-sister—will become embryos. They are no different from any other sibling combination except that they were nurtured in the womb at the same time rather than, typically, a few years apart. Fraternal twins are often referred to as double-egg twins, and these are the types of twins that tend to run in families.

Identical Twins

These two, on the other hand, developed from the same egg and the same sperm and therefore have the same genetic makeup. How do two (or, in the case of triplets or quadruplets, three or four) identical children emerge from one egg? It is because a single fertilized egg, when it begins to divide by mitosis and grow, produces daughter cells identical to itself; rather than staying attached to one another in a single ball, they may split into two (or more) groups. Each group has the same genetic makeup, and each begins to grow at the same rate. Two groups of daughter cells produce identical twins, three groups produce identical triplets, and so on. Many mistakenly believe that identical twins can be male and female. Their identical genes clearly make this impossible. Moreover, this so-called "single-egg" twinning does not appear to run in families.

What many may not realize is that twins are not truly identical, even though they have just been described as such. If the twins implant in the uterus at different places, they will receive substances from the mother's bloodstream differently; one twin may be exposed to a greater concentration of certain bacteria than the other. Their genetic makeup can change slightly after the blastocyst splits into two groups, because a few of one twin's genes may be damaged during subsequent cell division while the other's are not. After birth, they are exposed to environmental influences that will affect the genetic makeup of each twin differently. Then there is genetic imprinting, the process by which certain genes are turned "on" or "off" based on which parent they came from; because this somewhat random gene activation will occur differently in each twin, their tissues—some imprinted, some not—will develop somewhat differently.

For this reason, many scientists prefer "monozygotic" to "identical," and nowhere is the reason for this preference better illustrated than in forensic investigation. If a monozygotic twin is suspected of a crime, DNA evidence cannot identify which twin is the culprit because current analytic techniques cannot discern the tiny mutations and variations that differentiate the two. So for now, investigators must hope the guilty twin left some fingerprints behind because, surprisingly, the fingertips of monozygotic twins are not the same. In an excellent example of how environment literally shapes development, the different whorl patterns of identical twins' fingerprints arise in part from the amniotic fluid surrounding them as well as their contact with each other, themselves, and the walls of the uterus.

---

the last menstrual period through the end of the 12th week. The second trimester is almost four months, covering the 13th week through the 27th; and the third occurs between the 28th week and the 40th. Another system for marking pregnancy's passage, used during the 10-week-long first trimester, pinpoints 23 stages of embryogenesis beginning with fertilization at Stage 1, quickly followed by cleavage at Stage 2.

Cleavage
Fertilized at the upper end of the Fallopian tube, the zygote must make its way to the uterus, implant, and begin developmental growth. As it migrates, it undergoes cleavage, or cell division. The resulting daughter cells divide

every few hours, creating a cluster that is the multicelled embryo. The cells then flatten to form tighter junctions, compacting into a rounded morula that enters the uterus about the fifth day after fertilization.

At Stage 3, the morula develops fluid at the center and becomes a blastocyst. As fluid accumulates, the blastocyst for the first time begins to display signs of cell differentiation. There is an inner cell mass that represents the embryo-to-be; a discernible outermost collection of tropho-blast cells that later contributes to the placenta, an organ that supplies nutrients to the fetus and removes waste products such as carbon dioxide; and the umbilical cord, a network of blood vessels that connects the embryo to the placenta. The fluid-filled center is contained by the amnion, a membrane surrounding the embryo that suspends it in amniotic fluid.

Identical twins can result from cleavage if daughter cells split and implant at different sites in the uterus, each developing its own amnion and placenta. These cases account for about 10 percent of identical twins. Another 70 percent or more are formed when the inner cell mass splits, forming two embryos and two amniotic sacs. Or, both may share an amniotic sac; if the cells fail to separate completely, the embryos will be conjoined (or Siamese) twins. Twins sharing an amniotic sac and placenta are at high risk.

Nearly a week after fertilization, the zygote is ready to implant in a uterus already prepped by the pituitary's hormonal directives. At Stage 4, the cluster of cells erupts from the still-intact zona pellucida to land on the blood-rich endometrium of the uterus, where it burrows aggressively and initiates physiological changes in its maternal host that she will soon be unable to ignore.

The Primitive Streak and Gastrulation

About this time, Stages 5 through 8, a placenta forms and becomes anch-ored to the uterine wall by means of chorionic villi, or "fingers," that infil-trate the endometrium. The blastocyst's inner cell mass divides into an epiblast that forms the embryo and a hypoblast that forms the yoke sac, a structure that supplies early nutrients in other animal embryos but may be vestigial in humans. A groove called the primitive streak forms along the back of the inner cell mass. Cells along this groove migrate inward in a process called gastrulation, during which they differentiate into three rudimentary tissues (embryonic germ layers) that in turn give rise to different organs and body systems: the outer ectoderm, cells that will form the nervous system, outer tissues like skin, hair, mouth, and anus, and the lenses of the eyes; the endoderm (sometimes called the gastroderm), an innermost layer that will become the linings of glands and internal organs comprising the digestive, respiratory, and endocrine systems; and the middle layer known as the mesoderm that will develop into the reproductive system, heart, lungs, and blood, and whose cells will develop segmented tissues called somites to become muscles and bones. The notocord, a rodlike structure that orients the organism to top and bottom, front and back, left and right, and forms the basis of the backbone, will also emerge from the mesoderm, followed by primitive development of the nervous system in Stage 8.

Occurring close to the third week of embryogenesis, gastrulation is a pivotal point of development in the first trimester because it marks the beginning of organ formation. This is a critical time for the embryo, since certain viruses or drugs or inadequate nutrition can gravely imperil its development and is one of the many reasons prenatal care is so important early in pregnancy.

The Embryo
The embryo's first three months of life are characterized by the most dramatic changes a human organism ever experiences. With a developing notochord serving as an axis, the three cellular layers curl under themselves to form a tube with a hollow "gut" through the middle. The body plan is taking shape, and primitive blood vessel networks are developing. Already elongating, the embryo's lumpy body indicates accelerating tissue differentiation. Thickened circles along its upper sides suggest the eyes that are to come and, nearby, puckered areas reveal the sites of future ears. By Stage 9, its primitive heart begins to beat and blood vessels grow, even as a large forebrain starts to dominate the embryonic structure.

In Stages 10 through 12 (weeks 3 and 4), changes become more marked. The embryo curls into a "C" shape, displaying budlike limbs. Digestive organs and glands start to develop, and the early division of the heart into distinct sections begins when blood circulation is established and heart valves are more defined. Extensive groundwork for central nervous system development occurs during these stages. A thin layer of skin forms over facial features that are beginning to distinguish themselves; nasal pits are visible, and the depression that will become the mouth appears.

Stages 13 through 18, during weeks 5 through 7, are characterized by the appearance of a fully developed umbilical cord, lengthening appendages, and a lobed heart. Digestive organs can be seen now, especially the intestines, which have grown so quickly they temporarily reside outside the abdominal cavity. The esophagus develops out of the trachea, and the lungs form. The embryo's trunk straightens to hold its head more erect, and its bones become harder. Fingers and toes are becoming more distinguishable, and genital membranes primed for male or female development reveal the influence of the Wolffian and Müllerian ducts described earlier in this chapter.

An increase in brain size, the growth of vocal cords, and progression of organ development characterize the next three weeks of the first trimester, Stages 19 through 23. Male or female genitalia become distinct as the testes begin their slow descent into the scrotum or the ovaries relocate to the pelvic region. The embryo's chin sharpens. The pancreas begins to secrete insulin, hands display dexterity, and both body hair and fingernails grow. By the end of the eighth week, when its facial features are recognizably human, the embryo becomes a fetus (Figure 9.7). Only one inch (2.54 centimeters) long, it appears to swallow, even hiccup, and in its tiny mouth are 32 buds that will become permanent teeth. The muscles function more smoothly and the fetus makes random movements, although its mother cannot yet detect them.

The Mother

It is very unlikely the mother is having any symptoms of her pregnancy this early, but already her body is undergoing radical changes to accommodate a bundle of cells that her body regards as foreign—in this case, the embryo, whose genetic material is different from hers. Human immune systems do not accept alien invasions gracefully; they make very sharp distinctions between "self" and "nonself," and the mother's system would immediately identify this stranger as "nonself."

But just as her system poses a problem, it provides a solution. First, before the fertilized ovum descends all the way down the Fallopian tube,

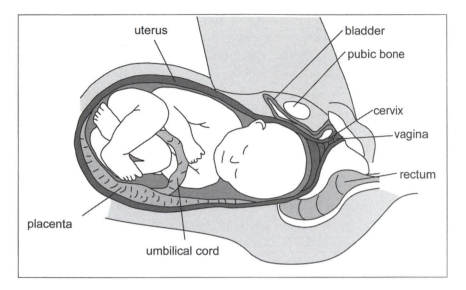

Figure 9.7 Full-term fetus. (Sandy Windelspecht/Ricochet Productions)

the mother's uterine mucus lining forms a membrane called the decidua so that, when the embryo implants, the decidua wraps over the organism to separate it from the inner wall of the uterus. This prevents the mother's immune system from "seeing" the embryo. Second, because there must be some kind of connection between her uterus and her child if the embryo is to survive, a placenta begins to develop from both the decidua and the fetal chorion, a layer of tissue that lies between the embryo and the decidua. This becomes the critical passageway for the transfer of nutrients and wastes between mother and child. Life-sustaining chemicals pass back and forth between the two: from the embryo to and from the placenta via the umbilical cord, and from the mother to and from the placenta via its attachment to her uterine wall. This vital activity escapes the surveillance of the immune system because the blood systems of mother and embryo do not touch. How, then, do they make the exchange? There are complex theories that genetic imprinting and graft rejection molecules are involved, but to what degree is not known. What is known is that the biochemicals exchanged by mother and fetus rely on the placenta, where osmosis and diffusion make the transfer.

The placenta also produces hormones, initially just enough to prevent the degeneration of the ovary's corpus luteum that is keeping the endometrial lining richly supplied with blood. When the placenta is capable of producing enough hormones on its own, usually after the first three months, the corpus luteum begins to disintegrate. The yolk sac usually disappears by the seventh or eighth week.

Much of the embryo's very early development occurs before the mother even realizes she is pregnant. If she becomes aware of anything unusual, it might be that her menstrual period is a day or two late, but even in this she could be misled by the light vaginal bleeding or "spotting" that many pregnant women experience during embryonic implantation. (To add to the confusion, some women actually continue to have periods throughout their pregnancy, but this is rare.) Her breasts may be sore too, the kind of tenderness that just precedes her period. So, unless she is consciously awaiting confirmation that she is pregnant, she is oblivious to her condition.

That will change by the time her period is four weeks late. The embryo's primitive heart has been beating for two weeks, and the mother will have developed symptoms such as fatigue, lightheadedness verging on fainting, and "morning sickness." The latter is a misnomer; the syndrome is more properly known as "pregnancy sickness," or the "nausea and vomiting of pregnancy" (NVP) because the nausea can occur anytime, even in the middle of the night, and may range from slight queasiness to persistent vomiting and dehydration. If the nausea is severe (hyperemesis gravidarum), medical intervention is required, but for the usually mild-to-moderate sickness associated with the first trimester of normal pregnancy, small but frequent meals, avoidance of certain foods, and plenty of fluids will often alleviate symptoms. Although the causes of NVP are not fully understood, the most common hypothesis points to elevated hormones. The reasoning is that higher hormone levels seem to support a healthier pregnancy; women with healthier pregnancies—those who miscarry less frequently—have more NVP. Others believe that NVP is an evolutionary adaptation to prevent mothers from eating foods that are potentially toxic to the embryo.

The high levels of hormones a pregnant woman produces are due in part to the corpus luteum producing more estrogen and progesterone by order of the beta human chorionic gonadotropin (ß-hCG or hCG) hormone. The placenta produces hCG once the embryo embeds in the uterine wall, and the hormone's increased level is one measure by which pregnancy is confirmed. The additional estrogen, which incidentally causes the mother's breasts to become sore, is needed in the development of the embryo's placenta, bones, and sexual characteristics. Progesterone, initially produced by the corpus luteum to build up the endometrial lining before ovulation, normally decreases after ovulation, so the lining will disintegrate and be expelled in menstrual fluid. In pregnancy, however, the uterine lining must be sustained, so the progesterone released by the stimulus of the corpus luteum, which in turn is nudged by hCG, ensures that the hormone levels remain high. The woman continues to produce hCG throughout her pregnancy at sufficient levels to prevent uterine contractions, to stimulate breast tissue growth, and to help support the placenta. She also produces human placental lactogen (hPL), both to help her breasts prepare for nursing and to regulate her insulin levels to compensate for some of the nutritional demands the baby is making. As the mother approaches her second trimester, the fetus, whose basic organs systems are now in place, will begin to make rapid gains in size and weight.

The Second and Third Trimesters

The Fetus

As the fetus lengthens through the torso, its physical features are melding into position; its eyes are moving forward, and its ears are close to their permanent location. It can hear, and its reflexes have matured. As the weeks pass and it continues to strengthen, its limbs and torso grow larger to achieve proportion with its head. It is covered in a downy coating called lanugo that it will shed before birth. Its heart is pumping more blood, and its limbs move rapidly.

Between the sixth and seventh months, the fetus has immature lungs, and only with very sophisticated support could it survive outside the womb, but as it develops the capacity to breathe on its own, its rapidly growing brain takes over many of its body's functions. Its testes, if it is male, or ovaries if female, are fully formed, and hair is growing on its head. It is so large now it must bring its knees up into the fetal position to have room in its mother's uterus.

By the ninth month, the fetal lungs have matured and it has built up significant body fat under its skin, which has thickened. Its gastrointestinal system, though immature, is fully functioning, and its nervous system reacts reflexively. Toward the end of the month, when it weighs between 6 and 9 pounds (4.1 and 5.7 kilograms) and is about 20 inches (50.8 centimeters) long, it is ready to be born.

The Mother

A pregnant woman usually feels better during the second trimester than during the first. Nausea and fatigue often vanish early in the fourth month, and she may be suffused with a sense of contentment and well-being. But she may also begin to feel clumsy and overweight.

Growing pressure compresses her bladder, making her urinate frequently; she sometimes has backaches; hormonal changes and the pressure from the baby's growth may have made her constipated and given her hemorrhoids. She might have noticed swollen veins in her legs and stretch marks over her abdomen and breasts.

Colostrum, the nutrient her breasts produce for the child's first postnatal day or two, leaks from her breasts, which are sore and tender. Unlike the nausea of the first trimester, she might have food cravings that she is compelled to satisfy even though she knows she must not gain too much more weight. Headaches are sometimes a problem, as is indigestion (acid reflux) due to the pressure of the baby pushing against the valve that separates her stomach and esophagus. Her uterus is so large now it displaces her diaphragm, making her short of breath, and she has frequent pelvic pain as her bones and joints withstand the growing baby's activity in her womb.

All of these symptoms are normal, although not everyone experiences them. But there are other, more serious, conditions that can arise during pregnancy that require careful medical attention, and this is why prenatal care to help avert them is so important.

Medical Issues Surrounding Pregnancy

Prenatal Care

Any woman contemplating pregnancy should be sure she is in good health, both for her own well-being and for that of the child. She should quit smoking, drinking alcohol, and taking illegal or prescription drugs (unless the latter are approved by her physician) for at least three months before she conceives.

If the father is a smoker, he should quit smoking as well to prevent second-hand smoke from endangering the infant.

Once she knows she is pregnant, a woman should see her physician or arrange an appointment at a clinic where health care personnel will take her medical history, perform a physical examination, and analyze her urine and blood to determine the state of her overall health. Specific information they seek includes her blood type, Rhesus factor (see "Complications of Pregnancy" below), the presence of infection, anemia, or sexually transmitted diseases, and whether she is immune to the German measles (rubella) virus that endangers normal fetal development. Other risks to the fetus about which pregnant women should be informed include

- The dangers of smoking

- Fetal alcohol syndrome

- Parasitic, bacterial, and viral infections, such as toxoplasmosis, puerperal fever, or chickenpox

- Dietary insufficiencies

The physician or midwife who oversees the pregnancy will set up regular appointments in which to check the mother's weight, blood pressure, and urine, and to feel her abdomen to evaluate the progression of the pregnancy. Under normal conditions, only one or two visits are required during the first trimester. They increase to one visit every four weeks in the second trimester and to one visit every two weeks in the third. From the 14th week on, the baby's heartbeat will be monitored as well, and its size checked from about the 20th week on.

Maternal and Fetal Testing

Several screening and diagnostic procedures are recommended at various stages throughout pregnancy to evaluate the progress of mother and child. Some are routine, and others are repeated or introduced when there is reason for concern.

- The ultrasound scan is routine and is normally done around 12 weeks to screen for Down syndrome and again at 18 to 20 weeks to verify fetal growth is occurring normally. If the scan captures a view of the external genitalia, it reveals the child's gender.

- A multiple marker test done at 15 to 18 weeks consists of blood analysis. Additional screening tests are performed in the first and second trimesters if specific complications or disorders are suspected based on the mother's history, symptoms, or the results of the ultrasound; these also detect chromosomal abnormalities and alpha-fetoprotein levels, which can be implicated in Down syndrome or neurological disorders.

- A glucose tolerance test will probably be administered to check for gestational diabetes.

- Amniocentesis is a diagnostic procedure usually performed only on mothers age 37 or older because of the small risk of miscarriage it poses. A sample of amniotic fluid is removed from the uterus and examined for fetal cells. These can yield valuable information like the sex of the fetus, its metabolic health, and whether there are chromosomal irregularities indicative of Down syndrome or other disabilities.

- Chorionic villus sampling is a less common diagnostic procedure than amniocentesis but can be done at an earlier stage of fetal development to detect genetic problems or blood abnormalities. This test is normally performed only if specific disorders are suspected.

- Umbilical vein sampling or cordocentesis elicits a great deal of information from fetal blood such as biochemical imbalances that may lead to slowed or retarded development, the presence of infection, and the Rhesus factor. If necessary, the umbilical cord can serve as the route of intrauterine blood transfusion.

Complications of Pregnancy

Despite early detection capabilities and the remarkable sophistication of today's screening and diagnostic tools, not all the complications associated with pregnancy can be avoided, nor can they be successfully treated once they pose a threat.

- Miscarriage and stillbirth are the death of the fetus. Miscarriages usually occur in the first trimester, often because of embryonic abnormalities too severe to allow survival. Stillbirth refers to death of the fetus after 24 weeks. Sometimes hormones fail to support the pregnancy, sometimes infections or illnesses compromise it, and sometimes the mother's uterus or cervix have impairments or irregularities that do not allow pregnancy to continue. Although nearly 25 percent of women experience some vaginal bleeding early in pregnancy, only half of those are likely to miscarry. Bleeding in the second and third trimesters, however, can be very serious and may indicate the placenta is involved.

- Placental separation occurs when the placenta becomes detached from the uterine wall, and the condition can have very severe consequences. In the first or second trimester, it means almost certain death of the fetus; in the third trimester, if the extent of the separation is not so great as to place the mother's life in danger from excessive bleeding, a Caesarean section can usually rescue the fetus.

- Placenta previa is a life-threatening development for the fetus because the placenta lies in the path of the baby through the birth canal, and if it is dislodged during the child's passage, it will cease to provide needed oxygen. Fortunately, the condition can be diagnosed early and, if the mother receives proper bed rest and specialized medical supervision up to her 37th week, she could deliver a healthy baby by Caesarean section.

- Placental insufficiency is just that; for unknown reasons, the placenta is occasionally unable to transfer nutrients and waste products with enough efficiency to sustain the fetus. Ultrasound imaging and detecting less-than-normal maternal weight gain can lead to a diagnosis of this condition, which may require the induction of labor or a Caesarean section.

- Preeclampsia and eclampsia are very serious conditions with unknown causes. They arise from the placenta and are somehow related to an inadequate blood supply. Preeclampsia, which is symptomless, is detected with blood pressure readings and urine tests during prenatal checkups. It rarely occurs before the 20th week, but when it does, the only treatment beyond stabilizing the mother is delivery of the fetus and placenta. If preeclampsia develops near term and the mother receives expert medical attention, she may be able to deliver her baby normally. In any event, the condition must be monitored constantly because it could suddenly lead to eclampsia, a worsening of the condition that results in seizures, kidney failure, coma, and, if not treated immediately as the medical emergency it is, death of the baby.

• An ectopic pregnancy means the embryo has implanted in a Fallopian tube where it promptly begins to grow. The discomfort and pain this causes usually alerts the mother by the 6th to 10th week of her pregnancy that something is very wrong. This is the subacute form of ectopic pregnancy in which the Fallopian tube is still intact and the embryo can be destroyed by an injection. The acute form, by contrast, comes on suddenly and is an extreme emergency. The tube ruptures and severe pain, shock, and falling blood pressure imperil the life of the mother. Immediate surgery to remove the tube and the fetus can save her life, but she may have difficulty conceiving again or may even be left infertile.

• Gestational diabetes is diabetes that begins in pregnancy and, in many cases, disappears after the baby is born. Diabetic mothers require careful monitoring, because maintaining proper blood sugar levels is critical to fetal health. Often they must deliver by Caesarean section because their children tend to be large; this is due to the increased sugar in the mother's system that crosses the placental barrier and becomes converted in the fetus into larger organs. With careful management, mothers who develop the disease during pregnancy should expect to deliver healthy children.

• The Rhesus factor (Rh factor) is simply an antigen or protein-like substance that appears in the red blood cells of about 85 percent of people; this makes them Rh positive. The other 15 percent of people are missing the antigen and are Rh negative. It becomes important when the blood of an Rh-negative mother mixes with that of her Rh-positive baby during delivery. No harm is done at that time except that the mother's immune system reacts to the presence of the baby's Rh-positive blood by producing anti-Rh-positive antibodies. If she gets pregnant again with an Rh-positive baby, the antibodies her system has produced will seek out and begin to destroy the baby's red blood cells. Monitoring the mother's blood for antibodies during the pregnancy allows physicians to evaluate the fetus and, if it is in distress, transfuse it through the umbilical cord.

- A Group B strep test is usually administered at the 35th to 37th week to determine whether the mother is carrying an infection that could be transmitted to her baby. She will also be examined to determine whether the baby is in the normal (head down) or breech (buttocks down) position for delivery, and may have her cervix checked to see if it has dilated in preparation for childbirth.

Preventing and Terminating Pregnancy

Since humans first made the connection between sex and pregnancy, they have sought ways to have the pleasure of the former without the consequence of the latter. They have often succeeded, although there were millions of unplanned pregnancies along the way. That does not have to be the case now, because there are several birth control methods that, if used properly, are nearly 100 percent effective in preventing pregnancy. Disease is another matter, however. Only condoms can help prevent sexually transmitted diseases.

Given the differing viewpoints about when life begins, it is important to distinguish between contraceptives and abortifacients. Contraceptives are agents that prevent ovulation, kill sperm, or block it from reaching the ovum. Abortifacients are agents that intervene after fertilization to prevent implantation of the blastocyst in the uterus and cause the embryo to be aborted. For this reason, contraceptives and combinations of contraceptives and abortifacients are discussed separately in the following sections.

If an unwanted pregnancy does occur, millions of women turn to abortion. The U.S. Supreme Court has ruled that abortion is legal up to the 24th week of pregnancy, but many states have imposed limits on the procedure. Some require parental notification or consent for minors, others require waiting periods, and still others have pushed the date of fetal viability to earlier than 24 weeks. New laws restricting abortion and court challenges to that legislation are pending in many states. Because abortion is defined as the premature delivery of a human embryo or fetus that cannot survive outside the womb, the use of abortifacients is sometimes referred to as abortion.

Contraceptives

The famous—or infamous—withdrawal method, or coitus interruptus, is hardly a birth control method at all because it fails so frequently. It amounts to the male withdrawing from the vagina immediately before he ejaculates. But even if he does, the lubricating fluids his glands produce during sex can be loaded with sperm well before he ejaculates.

The rhythm method is a natural birth control measure endorsed by the Catholic Church. (Other natural contraceptive practices like the body temperature method, described in the following paragraph, are presumably allowed as well, but the Catholic Church specifically bans any artificial means of contraception.) The rhythm method relies on the partners' avoidance of intercourse at exactly the time of month the ovum would be poised for fertilization in the Fallopian tube. Its effectiveness, if very carefully timed, approaches 80 percent. Like Catholicism, many religions, such as Orthodox Judaism and very traditional factions within Hinduism, Islam, and Christianity, impose restrictions on birth control except in certain cases, while some modern branches of these religions favor a more liberal approach to contraceptive use. In many cultures, opponents of contraception view it as a license for immorality, while proponents view it as essential to worldwide health and avoidance of many threats posed by overpopulation.

The Billings method and body temperature method are techniques in which a woman relies on her body to tell her when she is most fertile. In the Billings method, she carefully observes her vaginal discharge for a few critical days, watching for sticky, opaque cervical mucus that tells her she is ovulating. Some women may need informal training to be sure they can recognize the characteristic discharge. In the body temperature method, she simply takes her temperature during those same critical days to discern the slight elevation that occurs right after ovulation. These methods at best are 80 percent effective.

The cervical cap and diaphragm are "barrier" methods in that they block the entry of sperm into the uterus. The cap fits snugly over the cervix itself; the diaphragm is larger and is placed in the upper vagina in front of the cervix. Both devices must be fitted by a doctor, and both should be used with an over-the-counter spermicide to improve their effectiveness. They must be kept in place for several hours after intercourse, then removed and cleaned for reuse before having intercourse again. They are inexpensive and safe to use, although some spermicides irritate genital tissue, and the flexible ring surrounding the diaphragm can put enough pressure on the bladder to cause a uterine infection. Even with proper use, these are only about 80 percent effective, although that rate rises a few points if spermicides are conscientiously used as well.

A new contraceptive technique now on the market is a tiny springlike device implanted into the Fallopian tube near its entrance into the uterus. This blocks sperm from reaching the ovum where fertilization can occur. Considered permanent contraception, it cannot be reversed, so it is primarily aimed at older women who have had their families and want no more children. One advantage is that, unlike tubal ligation, the device is inserted vaginally through the uterus in a physician's office under local anesthesia. It requires no surgery, and recovery from the procedure is almost immediate.

Condoms are thin, flexible sheaths that fit over the erect penis and catch the sperm during ejaculation, blocking it from reaching the cervix. Condoms are effective only if they do not leak and should be checked carefully for cracks or holes before use. They are available in pharmacies and grocery stores everywhere and are often manufactured with a spermicide lubricant or packaged with a printed recommendation to use a spermicide for greater effectiveness.

So far, there are no other male contraceptives on the market except the condom, but research trials are underway to develop a hormone-based agent that suppresses sperm production.

There is also a female condom, a pouch that fits inside the vagina and performs the same function as the male condom. Male condoms are slightly more effective than female, 85 percent versus about 80 percent, respectively.

The spermicidal sponge, foam, cream, jelly, and suppositories are different products with the same mode of action. Any one of them can be inserted into the vagina before intercourse. They are safe and comfortable to use unless either partner is allergic to an ingredient in the spermicide. The sponge must be removed and disposed of after intercourse. Effectiveness can range from 80 percent to 85 percent, but if a woman has been pregnant before, the effectiveness of the sponge is reduced to about 60 percent.

Contraceptive-Abortifacient Combinations

The contraceptive pill, or "the pill" as it is generally known, created a revolution in birth control in the 1960s. It gave women complete control over their reproductive destiny for the first time and, many believe, spawned the sexual revolution of the 1960s by freeing women to explore their sexuality exempt from pregnancy. In the 45-plus years since, the pill's estrogen and progestin levels have been decreased to deliver effective contraception with minimal hormones. It works by suppressing ovulation, but it has a backup mechanism, in case the first fails, that irritates the uterine lining to prevent implantation of the embryo. Taken cyclically, it is fairly safe, but risks increase greatly for older women and those who smoke. It is about 95 percent effective.

While the pill must be taken daily, implants are small rods or capsules placed underneath the skin of the upper arm that are fully effective 24 hours after insertion. Like the pill, they deliver estrogen and progesterone to suppress oocyte development, prevent implantation, and thicken cervical mucus to interfere with the penetration of sperm. The implants, 95 percent to 99 percent effective in preventing pregnancy, can remain in place for three to five years, at which time they must be replaced.

In recent years, a new birth control pill was introduced that reduces a woman's menstrual period from once a month to once every three months. Although its mode of action is similar to that of the original pill, the active ingredient in the new product is taken for 84 days rather than 20.

Injections given periodically at a doctor's office are convenient, but can cause irritation at the site of injection. Depending on their formulation, they give one to three months' protection that is 99 percent effective.

Abortifacients

The "mini-pill" is another kind of birth control pill, but has only one hormone, progesterone, which inflames the uterine lining and makes it inhospitable to the fertilized ovum. It is about 88 percent to 99 percent effective.

The intrauterine device (IUD) or intrauterine system (IUS) is a small device that a doctor inserts into the uterus. Although some IUDs are treated with a spermicide, their primary mode of action is to irritate the uterine lining. They last for up to five years, are very affordable, and are effective up to 99 percent, but they can cause cramping and heavy bleeding.

The vaginal ring, inserted by the user into the vagina for three weeks, releases hormones that confer 95 percent to 99 percent effectiveness. The ring can be awkward to remove.

The patch, which may cause skin irritation where it is applied on the body, is a hormonal delivery system changed weekly, with three weeks on and one off. It is 95 percent to 99 percent effective.

Significant research is underway to develop a contraceptive vaccine, renewable with "boosters," that harnesses the immune system. Studies are particularly focused on producing antibodies that block the hormones supporting pregnancy or that make sperm and eggs resistant to fusing at fertilization. Much more research is needed to establish the efficacy of these agents and to confirm that normal fertility returns after the antibodies diminish in the body.

Surgical Abortions

In the first trimester, the most frequent procedure is vacuum abortion performed in a doctor's office or clinic in a single visit. The doctor widens the cervix by inserting a series of tapered rods, then suctions out the contents of the uterus. To prevent damaging the uterus, this procedure should not be performed before the sixth week of pregnancy.

In the second trimester, pregnancy may be terminated by inducing labor. Known as the induction or instillation method, an injection is given in a hospital setting that will cause the pregnant woman to go into labor and expel the fetus a few hours later. Another procedure, usually done with anesthesia on an outpatient basis, is a dilation and evacuation (D&E). This is similar to the aspiration method except that the cervical dilation must be greater and the contents of the uterus, which are larger, must not only be suctioned but scraped out with a curette as well. A similar procedure is dilation and curettage (D&C), which may or may not involve suctioning.

Abortion is rarely performed in the third trimester, and then only when the mother has a severe medical problem that prevents her from carrying the child to term, or when there is a gross fetal abnormality that would not permit the child to live. When an abortion must be performed at this stage, it is one of two kinds: dilation and extraction (D&X) or hysterotomy. The former is a partial-birth abortion, so called because the dead fetus is delivered vaginally. Hysterotomy, the alternative procedure, is very similar to a Caesarean section.

Although there are a few mild side effects associated with the contraceptives listed here, the contraceptive-abortifacients and the abortifacients, because they are formulated with hormones, can have side effects ranging from mild to serious and can pose significant risk to certain women. The IUDs and IUSs, because they involve devices that are placed inside the uterus, can cause infection and, if they perforate other organs, serious bleeding and damage. Anyone considering these birth control measures should educate herself about these side effects and risks and discuss her concerns with her healthcare professionals.

Summary

Throughout this encyclopedia, the interaction of all the systems has been emphasized. Each of the systems cannot function properly with the other. While the reproductive system is no different, it is the only system whose primary purpose is the continuation of the species. Sexual reproduction represents not only mixing the gene pool to pass along the genetic material that determines characteristics, but also imparting resistance to mutations that can threaten the species.

Females and males each have a distinct set of reproduction organs—an internal and external set. The female organs produce an egg, which results in pregnancy if successfully fertilized by the sperm produced by the male reproductive organs. However, there are methods to avoid pregnancy—some more successful than others. Currently, the most reliable forms of birth control are contraceptives, which include the birth control pill for women and condoms for men.

10

The Respiratory System

David Petechuk

Interesting Facts

- At rest, we breathe 15 to 20 times a minute and exchange nearly 17 fluid ounces (about 500 milliliters) of air with each complete breath in and out.

- Approximately 5 fluid ounces (about 150 milliliters) of the air we breathe in with each breath fills the passageways of the trachea, bronchi, and bronchioles.

- We breathe over 5,000 times a day, taking in enough air throughout a lifetime to fill 10 million balloons.

- The average set of human lungs has approximately 600 million alveoli (300 million per lung), creating a respiratory surface about the size of a singles tennis court or a square about 27 to 28 feet long on each side.

- At birth, an infant's lung is estimated to have approximately 20 to 30 million alveoli and 1,500 miles of airway passages.

- The right lung is slightly larger than the left.

- The capillaries in the lungs would extend 1,600 meters, or about one mile, if placed end to end.

- Every minute, 1.3 gallons (5 liters) of blood is pumped through the pulmonary capillaries and around the alveoli.

- Overall, blood takes approximately one second to pass through the lung capillaries, during which time it becomes nearly 100 percent saturated with oxygen, while losing all of its excess carbon dioxide.

- As a result of goblet cells and fine hair-like structures called cilia that help to filter foreign particles out of the air before they can enter the lungs, air breathed in through the nose is cleaner than air entering through the mouth.

Chapter Highlights

- Basics of the respiratory system, including major cells and components: nose and nasal cavity, pharynx, larynx, trachea, bronchi, alveoli, and lungs

- Development of respiratory system

- Respiration process

- How gases are transported

- Cellular respiration

- Respiratory diseases and disorders

Words to Watch For

Aerobic	Capillaries	Cytochromes
Allergies	Carbamino	Cytoplasm
Alveoli	compounds	Electron transport
Antibodies	Carotid arteries	system
Aorta	Carotid bodies	Epidemics
Aortic bodies	Chemoreceptors	Erythrocytes
Bohr effect	Chloride shift	Flavoproteins
Bronchi	Cilia	Gas exchange
Bronchioles	Conchae	Glycolysis

Goblet cells

Haldane effect

Hemes

Hyperventilation

Immune system
response

Inflammatory
mediators

Intercostal muscles

Macrophages

Mucociliary

Oxaloacetic acid

Oxidation-reduction
reaction

Oxidization

Oxygen dissociation
curve

Pandemics

Pathogens

Plasma

Pleura

Porphyrin

Respiration

Septum

Surfactant

Vagus nerve

Introduction

When the aging vaudeville entertainer Sophie Tucker was asked what the key to long life was, she replied, "Keep breathing." Although Tucker was making a joke, her answer was also correct in the most fundamental biological way. Although all of the human body's various systems are integral to life, none of them—from the cardiovascular to the nervous systems— would be able to function without the respiratory system. It is the respiratory system that garners the body's most basic fuel in the form of oxygen that we breathe in from the air. Every cell in our body uses oxygen to produce energy from food and drink. In fact, every chemical process throughout the body ultimately needs oxygen to take place. It is also through the respiratory system that the body eliminates carbon dioxide waste from cell metabolism. If the respiratory system ceases to function, death occurs within minutes as carbon dioxide rapidly reaches toxic levels in the blood.

When most people think of the respiratory system, they generally think of the relatively simple concept of breathing in and out, which is called respiration. But the respiratory system is a complex assemblage of organs and tissues that are integral to three different types of respiration. Breathing begins with nerve impulses that stimulate the breathing process, moving air into and out of the lungs through a series of passages from the nose down through the throat and into the lungs. Once the oxygen-rich air reaches the lungs, gas exchange (oxygen for carbon dioxide) occurs between the lungs and the blood. This process is called external respiration. Then, working in concert with the circulatory system, the now oxygen-rich blood is transported to all of the body's tissues where the gas exchange process occurs once again, this time between the blood and cells, with the blood passing oxygen into the cells and carrying away carbon dioxide to be eliminated via the lungs and expiration. This respiratory process is called internal respiration. Once the oxygen reaches the cells, it is used for a variety of specific energy-producing activities within the cells. This third form of respiration is called cellular respiration.

The respiratory system, especially the lungs, is unique from other systems in that it is in close and constant contact with the outside environment via the air we breathe. As a result, it is exposed to a wide variety of potentially harmful substances, from naturally occurring bacteria and viruses to pollutants produced by humans and modern society. Largely because of these exposures, respiratory diseases and illnesses—from the common cold to asthma to lung cancer—are among the most prevalent forms of sickness and disease in human beings.

This chapter provides an overview of the respiratory system, from the basic anatomy and functioning of the system, as well as touching briefly on related diseases and treatments. Breathing is an amazing and intricate process, and the respiratory system is the very foundation of life.

Components and Development

From the day we are born and take our first breath, we have set into motion the continuous and essential process of acquiring oxygen (O_2) from the air and eliminating carbon dioxide (CO_2) from the blood. This exchange of gases is called **respiration**. The spontaneous and rhythmic process of breathing is made possible by a complex, finely tuned system of organs, tissues, and passages called the respiratory system. Working in conjunction with the cardiovascular system, which pumps 1.3 gallons (5 liters) of blood through the lungs every minute, the respiratory system provides oxygen for the body's cells to produce energy and removes the carbon dioxide waste by-product created by cellular metabolism.

In addition to **gas exchange**, the respiratory system has other functions. For example, the respiratory tract is lined from the nasal cavities to its smallest branches within the lung with sticky, mucous-secreting cells. These cells help defend the body from environmental pollutants by trapping and eliminating dust, allergy-causing pollens, and other airborne particles. The respiratory system also helps to maintain the body's temperature from 97° to 100°F by releasing warm, moist air during exhalation, and it plays a part in balancing the blood's acid-base alkaline composition. Nevertheless, the system's primary function is respiration for gas exchange. There are two distinct modes of respiration: organismal (sometimes referred to as external) respiration (involving the lungs) and cellular respiration (involving chemical reactions within the cells). All of the respiratory system's functions begin with a specialized system of structures and organs.

The Components

The respiratory system can be broken down into two portions, each of which performs distinct functions. The conductive portions are composed of structures that act as ducts and pathways connecting the lungs to the outside environment. These include the nasal cavity, pharynx, and other structures. The respiratory portion, which includes the lung and lung structures, facilitates the gas exchange process. In addition, the respiratory system includes ventilating mechanisms, which are the various chest structures and muscles that help to move air in and out of the lungs. The entire respiratory system can also be broken down into two sections: the upper and lower respiratory tracts.

The primary components of the upper respiratory tract are:

• Nose and nasal cavity (passage)

• Pharynx (throat)

• Larynx (voice box)

The primary components of the lower respiratory tract are the:

• Trachea (windpipe)

• Bronchi

• Alveoli

• Lungs

The Nose and Nasal Passages

Although we sometimes breathe through our mouths (for example, when we run or do strenuous work, or when we have a sinus infection), human inspiration (taking in air) and expiration (expelling air) usually takes place through the nose and nasal cavity, which joins the nose and the pharynx. The nasal wall, or **septum**, divides the nasal cavity into two sides. The bottom portion of the nasal cavity is called the hard palate, and three bony ridges or projections, called nasal **conchae**, are on the surface of the cavity sides. The nasal structure also includes the paranasal sinuses. These hollow cavities in the bones of the head connect to the nasal airways via a small passageway in the conchae called a meatus. It is unclear what function the paranasal sinuses perform. They may help provide resonance for vocal sounds and lighten the skull. Another theory is that the sinuses may have once aided humans in the ability to smell, as they still do for some lower animals. However, since they no longer perform this function in humans, the paranasal sinuses may be a leftover component that no longer serves an important functional purpose.

Air enters through the external openings of the nose, called the nostrils or external nares. It then passes into the pharynx or throat through interior openings called the internal nares. The nasal passages and sinuses between the external and internal nares are lined with mucus-secreting epithelial cells called **goblet cells** and fine hair-like structures called **cilia** (see Sidebar 10.1). Together, these components help to filter foreign particles out of the air before these particles can enter the lungs. This filtering process is achieved when the sticky mucous membrane traps foreign particles, which are then swept by the waving microscopic cilia into the back of the throat, or pharynx, much like seaweed or sea grass buffeted back and forth by the waves. These particles are swallowed and eventually broken down by hydrochloric acid in the stomach and eliminated by the digestive system. The cough reflex can also expel them into the air. As a result of this process, air entering through the nose is cleaner than air entering through the mouth.

The nose and nasal cavities also serve as the body's air conditioner. The nasal passages and mucous membrane warm and humidify air before it enters the lower part of the respiratory system; this function is essential

SIDEBAR 10.1

Major Cells of the Respiratory System

Epithelial cells typically form sheets covering the surface of the body and lining cavities, tubular organs, and blood vessels. They play a major role in the respiratory system. Pseudostratified columnar epithelium cells line the conducting portion of the respiratory tract, from the trachea to the mid-size bronchioles. They are called pseudostratified columnar because this sheet of columnar cells (cells that are taller than they are wide) look like they are stratified in layers. However, the "pseudo" prefix means "fake," and these cells are not actually multilayered. Cells making up the pseudostratified columnar epithelium include:

- Ciliated cells that have moving cilia to "sweep up" particulate matter

- Goblet cells that produce and secrete mucous coverings (primarily in the trachea and bronchi), help humidify the air, and trap foreign particles

- Basal cells in the bronchi and bronchioles that may serve as stem, or progenitor, cells to create other cell types, including ciliated and goblet cells

- Clara cells that secrete extracellular lining fluid and surfactant proteins

In addition, two essential types of epithelial cells are found in the alveoli:

- Type I pneumocyte (Alveolar type I) cells are very thin, flat squamous cells that cover about 95 percent of the alveolar surface and form part of the blood-gas barrier for gas exchange in the alveoli.

- Type II pneumocyte (Alveolar type II) cells are situated at the junctions between alveoli and synthesize and secrete phospholipid-rich surfactant; they also proliferate in response to lung injury acting as a progenitor, or precursor, for the type I cells.

to help prevent harm to other, more fragile linings within the system, such as the lining of the lungs. Several features facilitate this process. Humidification takes place partially because of moisture secreted by the mucous membrane. The nose is also partitioned into two halves by the nasal septum, which is supported by bone and cartilage, thus providing a greater surface area for warming air. **Capillaries** (small blood vessels) that line the nose and cavities also give off heat, and the nasal conchae folds increase surface area and create turbulence that further "conditions" the air.

Pharynx

The pharynx, commonly referred to as the throat, is the funnel-shaped opening leading from the nose and mouth to both the lower respiratory tract and the digestive system. While food passes through the pharynx into the esophagus and stomach for digestion, air passes through the nose and pharynx and on into the larynx and trachea, which leads directly into the lungs. The pharynx, which is about five inches long, is typically divided into three segments, called the nasopharynx (upper), oropharynx (middle), and laryngopharynx (lower). (The nasopharynx serves exclusively as part of the respiratory tract. The oropharynx and laryngopharynx also help guide food into the alimentary tract. On swallowing, a muscular flap called the soft palate closes the nasopharynx off from the oropharynx. The laryngopharynx connects the oropharynx with the esophagus.) Lined with mucous secreting epithelial cells to help remove foreign particles, the pharynx also helps to warm and humidify air before it reaches the lungs.

Larynx

Composed of bone, cartilage, and muscle, the larynx is a valve-like structure that separates the trachea from the upper respiratory tract and connects the pharynx and trachea. It includes the large thyroid cartilage that can protrude from the front of the neck, commonly called the "Adam's apple." Although often referred to as the "voice box" because it contains the vestibular (vocal) folds and chords needed for human speech, the larynx serves important regulating functions during respiration. Both the vestibular folds and the epiglottis, a flap-like tissue composed of elastic cartilage that sits above the larynx, act similar to trap doors that open to allow air to enter and close to prevent aspiration (food from entering the lower respiratory tract). The larynx, which is also lined with mucosal epithelium, also helps the respiratory system rid itself of impurities through the coughing mechanism activated by nerves that are extremely sensitive to touch. Laryngitis develops when mucosal epithelium on the vocal chords become inflamed.

Trachea

The trachea, commonly referred to as the windpipe, is a tube-like structure stabilized by 15 to 20 C-shaped pieces of cartilage (Figure 10.1). It is typically 4 to 5 inches (10 to 12 centimeters) long and around one inch

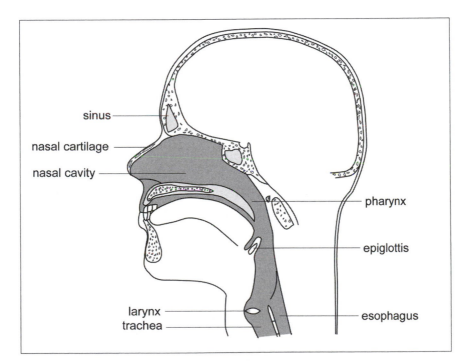

Figure 10.1   Upper Respiratory Tract. The upper respiratory tract, with the upper portion of the trachea and esophagus included. (Sandy Windelspecht/Ricochet Productions)

(2.5 centimeters) in diameter. In addition to serving as the primary passage-way of air into the lungs, the trachea contains mucus-producing epithelium to trap foreign particles. Cilia are also present to propel these particles upward toward the larynx for swallowing or expiration. As the lower end of the trachea enters the lungs, it branches off behind the sternum (breast-bone) into the left and right primary **bronchi**, which enter the left and right lung. Because the right bronchus is shorter, wider, and more vertical than the left bronchus, food usually enters the lower respiratory tract via the right bronchus when it bypasses the esophagus and "goes down the wrong pipe."

Bronchi and Bronchioles

The primary bronchi are similar to the trachea in that they also have an epithelium lining and are supported by C-shaped cartilage. The final por-tion of the respiratory system's conductive segment, the primary right and left bronchi branch off further into increasingly smaller bronchi down to approximately 0.04 inches (1 millimeter) in diameter. These differ in construction from the primary bronchi in that their support comes from smaller cartilage plates embedded in the walls.

The complex system of bronchi that branches throughout the lungs is called the bronchial tree, which extends further and further into finer "branches" that have less cartilage for support and more smooth muscle. These ultimately become **bronchioles**, which are approximately 0.02 inches (0.5 millimeters) in diameter. The division leads to terminal bron-chioles and then respiratory bronchioles. These tiny tubes, which further divide to form alveolar ducts that will end in air sacs called **alveoli**, are considered the first structures that belong to the respiratory portion rather than the conductive portion of the respiratory system.

Alveoli

The respiratory bronchioles end in small grape-like clusters of alveoli (individually called alveolus), where the gas exchange between oxygen and carbon dioxide takes place (Figure 10.2). As bronchioles continue to divide, the number of alveoli increases. The average set of human lungs has approximately 600 million alveoli (300 million per lung), creating a respiratory surface in the vicinity of 750 square feet (about 70 square

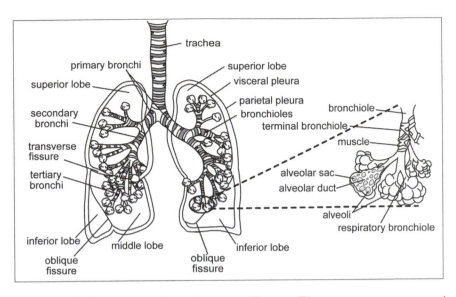

Figure 10.2   Lower Respiratory Tract. The major passages and structures of the lower respiratory tract. (Sandy Windelspecht/Ricochet Productions)

meters, which is about the size of a singles tennis court or a square about 27 to 28 feet long on each side).

Each individual alveolus is approximately 0.004 to 0.007 inches (0.1 to 0.2 millimeters) in diameter. Like tiny balloons, the alveoli inflate and deflate. Alveoli have thin, one-celled walls made of squamous epithelial cells (epithelial cells with a scaly outer layer) that are covered by an extensive network of fine capillaries. Each single alveolus is surrounded by about 2,000 segments of capillaries, which have single-layered endothelial cell walls. Gas exchange occurs via diffusion (net movement of particles from a region of higher concentration to a region of lower concentration) between the thin walls of the alveoli and the capillaries. The blood-gas barrier, or respiratory membrane, has a thickness of approximately one-half of one micrometer (a micrometer is 1/1000 the thickness of a dime). The process involves oxygen passing from alveoli into capillaries for distribution throughout the body, and carbon dioxide diffusing from the capillaries into alveoli where the gas is eliminated through expiration.

A fluid called a **surfactant** is produced by type II pneumocytes (specialized cells that line the alveoli) and secreted in the alveoli to coat the walls and reduce surface tension or stiffness. Reduction of surface tension results in less pressure being needed to inflate the alveoli, which is especially important at birth. Surfactant lining the alveoli also provides the moist surface necessary for gas exchange, because gas must dissolve in liquid before moving through cells. As a substance, surfactant has a half-life of 14 to 28 hours, meaning that it degrades very quickly and must be continually produced by the pneumocytes.

Lungs

The bronchial tree and alveoli course throughout the conical-shaped left and right lungs. In terms of volume, the lungs are one of the largest organs of the body, and the two lungs together weigh a total of approximately between 1.7 and 2.2 pounds (800 and 1,000 grams). They take up the majority of chest space (thorax), which comprises the space from the base of the neck to the diaphragm, upon which the lungs sit. The slightly larger right lung has three lobes (the superior, middle, and inferior lobes), and the left lung has two (the superior and inferior lobes). Deep fissures, or crevices, on the lung's surface define the separate lobes.

Each lung is enveloped by a transparent membrane called the **pleura**, which has an outer membrane (parietal pleura) attached to and lining the thoracic, or chest, wall and an inner membrane (visceral pleura) that tightly covers the lungs. Between the outer and inner pleural membranes, which are actually one continuous membrane that doubles back to cover both the chest and lungs, is a space called the pleural cavity. Inside this cavity is the pleural fluid, which helps to reduce friction between the membranes during breathing when the lungs expand and contract and also helps hold both pleural layers in place, much like two microscopic slides that are wet and stuck together. The lungs are also encased by the rib cage, which provides protection from outside trauma. Between the right and left lungs is an area called the mediastinum, which contains the heart, trachea, esophagus, thymus, and lymph nodes. The heart separates the right and left lung, and the left lung's smaller size includes a "cardiac notch" to provide space for the heart to extend into.

Diaphragm and Intercostal Muscles

The diaphragm and the **intercostal muscles** are the ventilating mechanisms, or muscles, that allow the lungs to bring in and expel large volumes of air. At rest, we breathe 15 to 20 times a minute and exchange nearly 17 fluid ounces (about 500 milliliters) of air with each complete breath in and out. The dome-shaped diaphragm is attached to the lower six ribs via a central tendon. The intercostal muscles line the rib cage, with the external intercostals running forward and downwards and the internal intercostals running upwards and back. Together, they form sheets that stretch between successive ribs. The diaphragm and intercostal muscles help the chest area and the lungs to expand and contract. During inspiration, the external intercostal muscles contract and lift the ribs up and out, and the diaphragm contracts. This process increases the size of the chest cavity and reduces air pressure inside the lungs compared to the air outside,

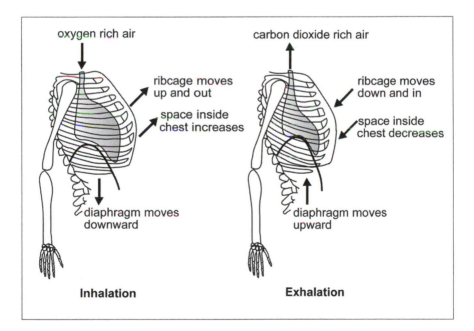

oxygen rich air

carbon dioxide rich air

ribcage moves up and out

ribcage moves down and in

space inside chest increases

space inside chest decreases

diaphragm moves downward

diaphragm moves upward

Inhalation

Exhalation

Figure 10.3   Inhalation and Exhalation. The movement of the diaphragm and rib cage during inhalation and exhalation. (Sandy Windelspecht/Ricochet Productions)

creating a vacuum that inflates and draws air (via the trachea) into the lungs. During expiration, the process is reversed, with the intercostal muscles moving the ribs downward and the diaphragm moving up creating a smaller chest cavity that increases lung pressure and forces air out. The natural elasticity of the lungs helps return them to their normal volume (Figure 10.3).

Development

The respiratory system begins to develop in the embryonic stage of life as cells start to divide. It continues postnatally (after birth) for at least two years and possibly for as long as 10 years. This multi-event process involves more than 40 different cell types that differentiate (specialize in structure and function) and proliferate.

Differentiated tissue for all the body systems and components develop from three primordial germ cell layers formed in the embryo during its early development. These are the ectoderm (outermost layer), mesoderm (middle layer), and endoderm (innermost layer) of the forming embryo. The respiratory system develops primarily from the mesoderm and endoderm. The endoderm germ layer differentiates into the larynx, trachea, and lung, and ultimately the lining of the respiratory tract. The mesoderm gives rise to the vascular system necessary for transportation of oxygen, as well as to other connective tissues, lymphatics, bone, and cartilage throughout the body.

Respiration

We are taught at an early age that eating the proper foods is essential for good health. But without the respiratory system, the food we eat could not sustain us. The oxygen (O_2) supplied by the respiratory system plays a fundamental role in enabling the body's cells to turn food into life-producing energy.

Beginning with nearly 2 gallons (approximately 6 to 7 liters) of fresh air we breathe in every minute, the respiratory system inhales close to 3,000 gallons of air each day to acquire the oxygen necessary to fuel the metabolic processes that create energy from the carbohydrates found in food. This energy enables the body's cells to multiply and function. The process also results in the generation of the waste product carbon dioxide (CO_2), which the respiratory system helps to eliminate. If we did not breathe in fresh oxygen, carbon dioxide would rapidly accumulate to toxic levels within the blood and result in death. As discussed earlier in this chapter, the process of bringing in oxygen from the atmosphere and eliminating carbon dioxide is called gas exchange.

Although the word respiration comes from Latin meaning "to breathe again," respiration is much more than merely "breathing" air in and out. In the human body, respiration encompasses many processes, from the readily perceptible act of breathing to the hidden, complex metabolic machinery that continuously works within the body's trillions of individual cells. In terms of human biology, respiration operates on two basic levels. The first is called organismal respiration and refers to the entire human body, or "organism," taking in oxygen from the environment and returning carbon dioxide to it. The second is called cellular respiration and encompasses the metabolic activities that occur when the body's cells use oxygen and food to generate energy and produce carbon dioxide.

Organismal Respiration

Organismal respiration involves four stages:

- *Pulmonary ventilation*: Movement of air in and out of the lungs

- *External respiration*: Gas exchange between the lungs and the blood

- *Internal respiration*: Gas exchange between the blood and the body's cells (tissues)

- *Transportation*: Movement of oxygen and carbon dioxide through the body via the blood

Pulmonary Ventilation and the Mechanics of Breathing

The respiratory system's function begins with the exchange of large volumes of air between the environment and the lungs via inspiration and expiration. This process is referred to as pulmonary ventilation. Although pulmonary ventilation primarily serves to bring oxygen into the body from the atmosphere and as the final stage of expelling carbon dioxide waste from the body, it is during this initial process that the respiratory system performs many of its functions secondary to gas exchange. Air is warmed and moistened as it enters the nose, which helps to maintain our body temperature. The respiratory system also filters out environmental pollutants, such as dust. For example, mucus secreted by goblet cells lining the airways and lungs traps particles, and then cilia sweep the mucus upwards from the throat for swallowing or expulsion via coughing. The respiratory system also helps to balance our body's acid, or pH, levels through its role in regulating the elimination of carbon dioxide. Control of pH is essential for the proper functioning of enzymes, proteins, and other biological processes.

As with all the major systems of the human body, the respiratory system works in conjunction with other major systems. In the case of pulmonary ventilation, the nervous system exerts initial control over the breathing process, including the rhythm, rate, and depth of breathing. The message centers in the brain that control rhythmic respiration of breathing in and out are located in the brain stem and are called the pons and the medulla oblongata. These autonomic (or automatic) brain centers are more primitive than parts of the brain located in the cortex, which give us control over our movements and thoughts.

Whether we think about it or not, the pons and the upper portion of the medulla automatically regulate our breathing, which is why we can still breathe while we sleep. However, we can exert voluntary control over our breathing when needed. For example, we can hold our breaths for a certain length of time under water, or consciously make ourselves breathe faster or slower. The part of the brain that allows conscious control of breathing is located in the cerebral cortex.

During automatic respiration, specific **neurons** in the medulla and pons send signals to motor neurons in the spinal cord, approximately 10 to 12 times each minute (Figure 10.4). These nerve cells, in turn, signal the diaphragm and intercostal muscles surrounding the thoracic cage to contract, thus expanding the rib cage and lungs within it. When the lungs expand, pressure within the lungs becomes lower than the pressure in the atmosphere, causing air to rush in through the conductive portion of the respiratory system (nose, pharynx, larynx, trachea, and bronchi) until full

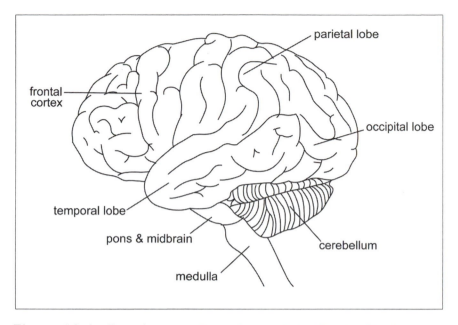

Figure 10.4 Respiratory Functions of the Brain. The location of the pons and the medulla, which are the parts of the brain that automatically control our rhythmic breathing. (Sandy Windelspecht/ Ricochet Productions)

expansion is reached. The air we inhale contains approximately 21 percent fresh oxygen and little or no carbon dioxide.

When the lungs are fully expanded, the **vagus nerve** tells the brain to "turn off" its signals for inspiration. As a result, the muscles surrounding the thoracic cage relax, moving the ribs back to their resting state. At this point, expiration is passive in that the brain and the muscles surrounding the lungs do not directly regulate it. The resulting decrease in chest cavity size contracts the lungs, causing us to exhale because the space available for air in the lungs is reduced. In other words, the air pressure in the lungs becomes higher than the pressure in the atmosphere. The resulting exhaled "stale" air is made up of about 16 percent oxygen and 6 percent carbon dioxide.

Although pulmonary ventilation is primarily automatic because we do not consciously control our breathing most of the time, the brain is actually hard at work interpreting **neurochemical** information to control breathing, including monitoring respiratory volume and blood gas levels. Surprisingly, the primary stimulus for the brain to control breathing is not the amount of oxygen in the blood, but rather the amount of carbon dioxide. Chemical receptors, or **chemoreceptors**, in the medulla, in collaboration with chemoreceptors in the **carotid arteries** and the **aorta**, respond to carbon dioxide levels in the blood. High carbon dioxide concentrations result in deeper and faster breathing designed to bring in higher levels of oxygen and reduce harmful carbon dioxide levels. In turn, our respiration rates slow down when carbon dioxide levels are lower. However, oxygen levels can also affect respiration. When the **aortic bodies** and **carotid bodies** detect low levels of oxygen, they send signals to the brain stem to make breathing more rapid and deeper.

Other factors affect the brain's regulatory function of respiration:

• An increase in blood pressure slows respiration.

• A sudden decrease in blood pressure increases respiration.

• A decrease in blood acidity (higher pH levels) increases respiration. (This state usually results from oxygen debt, or a lack of oxygen reaching the muscles, which produces lactic acid and lowers the pH level.)

Although some diseases can affect the body's oxygen and carbon dioxide levels, these levels are most commonly affected by physical activity. For example, hard work or exercise, especially when **aerobic** in nature, causes the body's cells to metabolize faster to create more energy. More carbon dioxide is produced in the process and eliminated into the blood, thus lowering the blood's **pH** level. As a result, during physical exertion, the body can increase oxygen consumption up to 25 to 30 times more than when the body is at rest.

External Respiration
External respiration is the exchange of oxygen and carbon dioxide between the lungs and circulating blood. Just as the respiratory system works in conjunction with the nervous system during pulmonary

ventilation, it also works with the heart and circulatory system to pump blood to the lungs during external respiration. This process is called pulmonary circulation. The movement of blood away from the lungs and heart to other parts of the body is called systemic circulation.

The amount of air we breathe in and out of the lungs during pulmonary ventilation is called tidal volume. Although we breathe in about one pint (around 500 milliliters) of air with each breath, approximately .32 pints (150 milliliters) of this air fills the passageways of the trachea, bronchi, and bronchioles. When filled with tidal volume air, these conductive portions of the respiratory system are referred to as anatomical dead space, meaning that the air remaining in this space is not involved in the external respiration process. The air that passes through the last conductive portions of the respiratory system enters the millions of alveoli in the lungs. It is here that external respiration takes place when oxygen and carbon dioxide are exchanged between air in the alveoli and the minute blood vessels called pulmonary capillaries that surround each individual alveolar sac like a net.

Every minute, 1.3 gallons (5 liters) of blood is pumped through the pulmonary capillaries and around the approximately 600 million alveoli in the lungs. The air-filled alveoli contain more oxygen compared to the blood in the capillaries. Conversely, blood in the capillaries contains more carbon dioxide than air in the alveoli. As a result, the exchange of oxygen and carbon dioxide between the capillaries and the alveoli occurs via diffusion across the microthin membrane walls separating the two.

The diffusion of all gases primarily depends on their solubility in water and their **partial pressure**, which expresses the concentration of a gas. The concentration, or diffusion gradient, of the gas is expressed in **millimeters of mercury** (mmHg). (For example, the partial pressure of oxygen would be expressed as pO_2 # mmHg.) Because the random movement of oxygen and carbon dioxide molecules results in their net movement from a region of higher concentration to a region of lower concentration, the concentration or partial pressure of oxygen in the alveoli must be kept at a higher level than in the blood. Likewise, the concentration of carbon dioxide in the alveoli must be kept at a lower level than in the blood. These different levels are maintained because the continuous inspiration of fresh air supplies an abundance of oxygen to the lungs and alveoli, while expiration eliminates carbon dioxide from the body into the air.

The blood carrying the carbon dioxide travels from all parts of the body into the heart's **right atrium** and then into the **right ventricle**, where it is pumped into the **pulmonary artery**. This artery, which is the only artery in the body that carries deoxygenated blood, branches into the right and left lung, ultimately feeding blood into the pulmonary capillaries. Blood entering the capillaries surrounding the alveoli has a pCO_2 of 45 mmHg and a pO_2 of 40 mmHg. Conversely, the environmental air that has entered the alveoli during inspiration has a pCO_2 of 40 mmHg and a pO_2 of 100 mmHg. As a result, oxygen diffuses across microthin membranes into the blood from the alveoli, and carbon dioxide diffuses into the alveoli from the blood.

When a person exhales, the air in the alveoli is breathed out into the atmosphere along with the abundance of carbon dioxide that it now contains. Conversely, the oxygen-rich blood is pumped throughout the body via the **systemic capillaries** to the cells that make up various tissues. Overall, blood takes approximately one second to pass through the lung capillaries, during which time it becomes nearly 100 percent saturated with oxygen while losing all of its excess carbon dioxide.

Internal Respiration
Although internal respiration is sometimes used in the same sense as cellular respiration to refer to the metabolic process within the cells, it is most often used to designate the gas exchange process between blood in the capillaries and the body's cells. Once the external respiration process is completed, the oxygenated blood travels from the alveoli to the heart's **left atrium**. The blood moves to the heart's **left ventricle**, then is pumped throughout the body via a network of arteries that feed the capillaries surrounding the body's various tissues. As mentioned earlier, this is known as systemic circulation.

When the blood returning from the lungs reaches the tissues, it has a pO_2 of 95–100 mmHg and a pCO_2 of 40 mmHg. Conversely, the cells that make up our tissues have a pO_2 of 30–40 mmHg and a pCO_2 of approximately 45 mmHg, depending on the metabolic activity within the cell. Again, since the diffusion of gases occurs from an area of higher

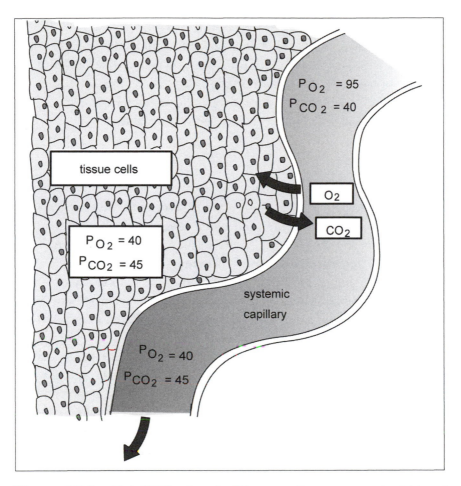

Figure 10.5 Gas Diffusion in Tissues. Oxygen-enriched blood (carried from the alveoli and the pulmonary capillary via the systemic capillaries) diffuses into tissue cells, which have a lower concentration of oxygen. At the same time, the higher concentration of carbon dioxide in the tissues diffuses into the systemic capillary for eventual transport to the alveoli. (Sandy Windelspecht/Ricochet Productions)

concentration to lesser concentration, oxygen from the blood diffuses across the interstitial fluid (liquid found between the cells of the body) and into the cells, or tissues. Conversely, the carbon dioxide from the cells and tissues diffuses into the blood (Figure 10.5).

Transporting the Gases

The gas exchange process necessary for cells to function properly could not occur without the blood transporting oxygen and carbon dioxide throughout the body. This transportation depends on the gases' distinct properties and on a blood component called hemoglobin, an oxygen carrying **protein** found in red blood cells called **erythrocytes**.

O₂ Transport

Compared to carbon dioxide, oxygen is not very soluble. As a result, only about 0.01 fluid ounces (0.3 milliliters) of oxygen will dissolve in every 3.4 fluid ounces (100 milliliters) of blood plasma, which is not enough to carry sufficient oxygen to the body's tissues and cells. The majority of oxygen in the human body is carried via hemoglobin, which is the respiratory pigment in humans and also gives blood its red color. Because of the affinity of oxygen to hemoglobin, the oxygen-carrying capacity of blood is boosted nearly 70-fold to about 0.7 fluid ounces (20.8 milliliters) per 3.4 fluid ounces (100 milliliters) of blood.

Hemoglobin's unique molecular characteristics make it an excellent transport molecule for oxygen. Each hemoglobin molecule includes four **hemes**, which are iron-containing **porphyrin** compounds, combined with the protein globin. Porphyrins are a group of organic pigments characterized by a ringed group of four linked nitrogen-containing molecular rings (called a tetrapyrrole nucleus). In a heme, each porphyrin ring has an atom of iron (Fe) at its center. Each iron atom can unite with one molecule of oxygen. As a result, each hemoglobin molecule can carry four oxygen molecules. Furthermore, when one oxygen molecule binds to one of the four heme groups, the other heme groups change shape ever so slightly so that their affinity increases for the binding of each subsequent oxygen molecule. In other words, after the first oxygen molecule is attached, the next three oxygen molecules attach even more rapidly to form oxyhemoglobin (the bright red hemoglobin that is a combination of hemoglobin and oxygen from the lungs), thus providing rapid transfer of oxygen throughout the blood. Conversely, when it comes time for hemoglobin to "unload" its oxygen content into cells and tissues, once one heme group releases its oxygen, the other three rapidly follow.

Oxygen's affinity for hemoglobin is also affected by the partial pressure of carbon dioxide and the blood's pH level. This is known as the **Bohr effect**, named after its discoverer Christian Bohr (1855–1911). A high concentration, or partial pressure, of carbon dioxide makes the blood more acidic, which causes hemoglobin to have less affinity for oxygen. As a result, in tissues in which the concentration of carbon dioxide in the blood is high because of its release as a waste product from cells, hemoglobin easily releases oxygen. In the lungs, where blood carbon dioxide levels are low because of its diffusion into the alveoli, hemoglobin readily accepts oxygen.

The Bohr effect or shift, which relates to a mathematically plotted curve called the **oxygen dissociation curve**, serves an extremely useful purpose. During exercise, cells are working harder—more actively respiring—to produce more energy. As a result, they release much higher levels of carbon dioxide into the blood than when the body is at rest. The higher carbon dioxide levels, in turn, reduce the blood's pH level, thus acidifying the blood and signaling hemoglobin to release more rapidly the oxygen needed to replenish cells and tissues. In other words, the Bohr effect informs the body that its metabolism has increased due to exercise and that it must compensate for the increased need to absorb oxygen and release carbon dioxide.

CO_2 Transport

Carbon dioxide enters the blood as a waste product of cell metabolism and cellular respiration. Unlike oxygen, carbon dioxide readily dissolves in blood. Carbon dioxide is transported by the blood to the alveoli in three ways:

1. As soluble CO_2 in blood (5–10 percent)

2. Bound by hemoglobin (20–30 percent)

3. As a bicarbonate (60–70 percent)

Although carbon dioxide is more soluble than oxygen and dissolves directly into the blood after it diffuses out of cells, the amount that dissolves is not enough to perform the essential function of ridding the body of carbon dioxide. In the second mode of transport, approximately a quarter of the carbon dioxide eliminated from cells reacts with hemoglobin. In essence, carbon dioxide is able to hitch a ride with hemoglobin because, at this point, hemoglobin is not carrying much oxygen and has an increased affinity for carbon dioxide. This is known as the **Haldane effect** and occurs as blood passes through the lungs. Blood proteins that bind to carbon dioxide are called **carbamino compounds**. When carbon dioxide binds to the hemoglobin's protein, the combination is called carbaminohemoglobin.

The first two methods of transporting carbon dioxide are relatively slow and inefficient compared to the third method of transporting the gas. Because carbon dioxide is highly soluble, it reacts readily with water (H_2O) molecules to form carbonic acid (H_2CO_3) in red blood cells. This reaction would also be too slow for efficient carbon dioxide transport if not for an **enzyme** called carbonic anhydrase (CA), which is highly concentrated in red blood cells and acts as a catalyst to help produce carbonic acid. The carbonic acid then ionizes (or disassociates) to form a positively charged hydrogen **ion** (H^+) and a negatively charged bicarbonate ion (HCO_3^-). The chemical process can be viewed as follows:

$$CO_2 + H_2O \longleftrightarrow H_2CO_3 \longleftrightarrow H^+ + HCO_3$$

Because the concentration of the negatively charged bicarbonate ions in the red blood cells is at a higher level than outside of these cells, these ions readily diffuse into the surrounding blood plasma for transport to the alveoli. To compensate for the negatively charged bicarbonate ions moving out of a red blood cell, a negatively charge chloride (Cl) ion enters the cell from the plasma to maintain the electrical balance in both the erythrocyte and the plasma. This exchange is called a **chloride shift**.

Cellular Respiration

Cellular respiration is the process by which cells use the oxygen delivered by the respiratory and circulatory systems to manufacture and release the chemical energy stored in food, primarily in the form of carbohydrates. As such, it is called an exergonic reaction, meaning that it produces energy. Cellular respiration produces energy via a catabolic process, that is, by making smaller things out of larger things. In cellular respiration, it refers to the breaking down of polymers (large molecules formed by the chemical linking of many smaller molecules) into smaller and more manageable molecules.

The catabolic process within cells involves breaking down glucose, a simple sugar in carbohydrates that stores energy, into smaller molecules called pyruvic acid. These smaller molecules are ultimately used to produce **adenosine triphosphate (ATP)**. ATP is the primary "energy currency" of the cell, the human body, and nearly all forms of life. Energy via ATP in cells is used to:

• Manufacture proteins

• Construct new organelles (subcellular structures that perform a role within each cell)

• Replicate DNA

• Synthesize fats and polysaccharides

• Pump water through cell membranes

• Contract muscles

• Conduct nerve impulses

Cellular respiration is the most efficient catabolic process known to exist in nature. Although it occurs in every cell in the body, cellular respiration does not take place simultaneously in the exact same phases throughout all the cells. If the energy produced though cellular respiration was released simultaneously, the body would not be able to process all the energy efficiently, which would result in wasted energy. In addition, the impact of such a large amount of energy being released all at once could overload and damage cells. As a result, cellular respiration occurs at different stages in the body's various cells, even in cells that are close neighbors or side by side. ATP molecules act like time-release capsules; they release small amounts of energy to fuel various functions within the body at different times.

Overall, two primary processes occur in cellular respiration. The first is the breakdown of glucose into carbon dioxide and hydrogen, known as the carbon pathway. The second is the transfer of hydrogen from sugar molecules to oxygen, resulting in the creation of water and energy. The entire process of cellular respiration occurs in three primary stages:

1. Glycolysis

2. Krebs cycle (citric acid cycle)

3. Electron transport system

Glycolysis

Glycolysis, which comes from the Greek words glykos ("sugar" or "sweet") and lysis ("splitting"), is the initial harvester of chemical energy within the body. It occurs in the cell's cytoplasm and converts glucose molecules into molecules of pyruvate, or pyruvic acid. Unlike the other processes in cellular respiration, glycolysis does not require oxygen and is the only metabolic pathway shared by all living organisms. Scientists believe that this biological approach to producing life-giving energy existed before oxygen developed in the Earth's atmosphere. It is the first step in both aerobic (oxygen) and anaerobic (oxygen-free) energy-producing processes (see Sidebar 10.2).

Glycolysis is a multistep process, with each step being catalyzed by a specific enzyme dissolved in the fluid portion of the **cytoplasm** called the cytosol. As with all biological processes, energy is needed to begin the process, and two ATP energy molecules initiate the reactions. This initial input of energy is called the energy investment phase, and occurs when ATP is used to phosphorylate, or add a phosphate to, the six-carbon glucose molecule. However, the process also yields energy in that further breaking down the six-carbon glucose molecule into two three-carbon pyruvic acid molecules ultimately results in a net gain of ATP molecules, as well as other energy molecules such as reduced nicotinamide adenine dinucleotide (NADH). However, glycolysis is extremely inefficient. The entire process captures only about 2 percent of the energy that is available in glucose for use by the body. Much more energy is available in the two molecules of pyruvic acid and NADH produced during glycolysis. It is this potential energy that goes on to the next step, called the Krebs cycle.

SIDEBAR 10.2

Anaerobic Respiration

When we exercise, our bodies produce more energy and require more oxygen. However, our blood cannot always supply enough of the oxygen via respiration that the cells in our muscles need. Under these circumstances, our muscle cells can respire anaerobically, that is, without oxygen, like some fungi and bacteria are able to do. Anaerobic respiration is also referred to as fermentation. However, cells in the human body can only respire without oxygen for a short period of time.

Like normal aerobic cellular respiration, anaerobic respiration begins with glucose in the cell, but takes place completely in the cell's cytoplasm. Although ATP energy molecules are also produced this way, the process is extremely inefficient compared to aerobic respiration. In the human body, the anaerobic process results in pyruvic acid being turned into the waste product lactic acid, as opposed to entering the mitochondria for further oxidation as it does in aerobic cellular respiration. It is the lactic acid in muscles that makes them stiff and sore after intense aerobic exercise, such as running.

The Krebs Cycle (Citric Acid Cycle)

Discovered by Hans Krebs (1900–1981), the Krebs cycle, also known as the citric acid cycle, is a cyclic series of molecular reactions that require oxygen to function. The cycle is mediated by enzymes that help create the molecules for the final harvesting of cellular energy in the third and final phase of the cellular respiration process. The Krebs cycle occurs in the matrix of the **mitochondria**, which are the powerhouses of cells. Although the mitochondrion is the second largest organelle in a cell after the nucleus, some cells may contain thousands of mitochondria from 0.5 to 1 micrometer in diameter. Unlike the energy-harvesting process of glycolysis in the cytoplasm, mitochondria are extremely efficient in taking energy from sugar (and other nutrients) and converting it into ATP. In fact, compared to the typical automobile engine, which only harvests about

25 percent of the energy available in gasoline to propel a car, the mitochondrion is more than twice as efficient—it converts 54 percent of the energy available in sugar into ATP.

After glycolysis is completed, the two pyruvate molecules that were formed enter the mitochondria for complete oxidization by a series of reactions mediated by various enzymes. As the pyruvate leaves the cytoplasm and enters a mitochondrion, acetyl coenzyme A (CoA) is produced when an enzyme removes carbon and oxygen molecules from each pyruvate molecule. This step is known as the transition reaction.

The Krebs cycle begins as oxygen within the cells is used to completely **oxidize** the acetyl CoA molecules. The process is initiated when each of the acetyl CoA molecules combines with **oxaloacetic acid** to produce a six-carbon citric acid molecule. Further oxidation eventually produces a four-carbon compound and carbon dioxide. The four-carbon compound is ultimately transformed back in oxaloacetic acid so that the cycle can begin again. Because two pyruvate molecules are transferred into the mitochondria for each glucose molecule, the cycle must be completed twice, once for each pyruvate molecule. Each cycle results in one molecule of ATP, two molecules of carbon dioxide, and eight hydrogen molecules. The ATP molecules produced during this cycle can be used as energy. But it is through the cycle's creation of the electron "carrier" **coenzyme** molecules NADH and reduced flavin adenine dinucleotide (FADH2)—which are created when the coenzymes nicotine adenine dinucleotide (NAD) and flavin adenine dinucleotide (FAD) "pick up" the hydrogen molecules—that the abundance of ATP is produced in the next stage of cellular respiration, the **electron transport system**.

Electron Transport System

Overall, the first two processes in cellular respiration, glycolysis and the Krebs cycle, have produced relatively little energy for the body's cells to use. Although both of these processes produce some ATP directly, the energy currency of ATP is created and cashed in for the big payoff during the electron transport system, also known as the electron transport chain. This process takes place across the inner membrane of the mitochondria called the cristae. A chain of electron receptors are embedded in the cristae, which are folded to create numerous inward, parallel, regularly spaced projections or ridges. This design results in an extremely high density of receptors, thus increasing the electron transport chain's efficiency.

The receptors are actually a network of proteins that can carry electrons and transfer them on down a protein chain. The process works like a snowball gaining speed as it rolls down a hill. As the NADH and FADH$_2$ molecules produced during glycolysis and the Krebs cycle pass down the chain, they release electrons to the first molecule in the chain and so on. Because each successive carrier in the chain is higher in electronegativity (that is, has a higher tendency to attract electrons) than the previous carrier, the electrons are "pulled downhill." During the process, hydrogen protons (H$^+$) or ions from NADH and FADH$_2$ are transferred along a group of closely related protein receptors that include **flavoproteins**, iron-sulfur proteins, quinones, and a group of proteins called **cytochromes**. The cytochrome proteins in the electron transport system will only accept the electron from each hydrogen and not the entire atom. The final cytochrome carrier in the chain transfers the electrons, which by this time have lost all their energy, to oxygen in the matrix to create the hydrogen-oxygen bond of water. This bond is another reason why oxygen is so important to the life of the cell. Without it, the molecules in the chain would remain stuck with electrons, and ATP would not be produced.

Because of the second law of thermodynamics, the electrons passed down the chain lose some of their energy with every transfer from cytochrome to cytochrome. Some of the energy lost helps to "pump" hydrogen ions out of the mitochondria's matrix into a confined intermembrane space between the mitochondria's inner and outer membranes. This energy for pumping the hydrogen ions is a result of a process called the **oxidation-reduction reaction**, or redox reaction. The reaction results in the molecules within the electron transport system alternately being reduced (gaining an electron) and then oxidized (losing an electron). The entire process establishes a buildup of hydrogen ions, resulting in a concentration, or diffusion, gradient—more hydrogen ions are pumped inside the confined space between the mitochondria's membranes than exist in the mitochondria's matrix. As the concentration gradient increases, the ions begin to diffuse back through the membrane into the matrix to equalize the hyrdogen ion gradient.

Hydrogen ion diffusion occurs through ATP synthase, an enzyme within the inner membrane of the mitochondrion. ATP synthase uses the potential energy of the proton gradient to synthesize the abundance of ATP out of the adenosine diphosphate (ADP) molecule and phosphate. This process is referred to as chemiosmosis. The formation of ATP is an energy storage process, and the energy is released when ATP is converted via the ATPase enzyme back into ADP (adenose bound to two phosphate groups) or to adenosine monophosphate (AMP—adenose bound to one phosphate group). All of these conversions are known as ATP phosphorylation. ADP and the separate phosphates produced by the breakdown are then recycled into cellular respiration for the recreation of ATP. At the same time, the waste products carbon dioxide and water are eliminated via diffusion from the cell into the bloodstream and on through the organismal respiratory process (Figure 10.6).

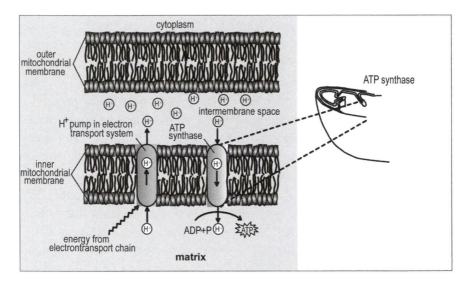

Figure 10.6   Synthesis of ATP. The buildup of hydrogen ions into the mitochondria's intermembrane space via electron transport and the eventual transport of these ions back through the membrane, where they are used by ATP synthase to make ATP (the major source of energy for cellular reactions) out of ADP and phosphate. (Sandy Windelspecht/ Ricochet Productions)

Respiratory Problems and Diseases

Few people have gone through life without experiencing an acute (short but severe) upper respiratory infection, more commonly known as the common cold. Not only are colds the most prevalent infectious disease known, respiratory ailments and diseases as a group are more common than any other medical problem in humans.

Numerous factors can adversely affect respiratory functioning, from genetic influences and medical problems during infancy to overall health as we grow older. Even psychological well-being may impact respiratory health. For example, some cases of bronchial asthma have been linked to anxiety, and a sudden anxiety attack can lead to **hyperventilation**. Nevertheless, environmental factors are by far the most common cause of respiratory ailments and diseases.

Because we breathe in approximately 16,000 quarts (over 15,000 liters) of air each day, our respiratory system is exposed to a continuous barrage of substances in the air that can affect the system's functioning, from bacteria and viruses to pollutants caused by industry and automobiles. According to some estimates, a person inhales and ingests approximately 10,000 microorganisms per day. Although the respiratory system is designed to protect our bodies against this environmental onslaught, it is not always successful, especially in cases of overexposure to pollutants. Cigarette smoking, for example, is directly responsible for the overwhelming majority of lung cancer and emphysema cases.

Respiratory System Defense Mechanisms

The respiratory system has several features that help protect it from the possible harmful effects of environmental particles and **pathogens** (viruses, bacteria, etc.) that can enter the system when we breathe. In the upper respiratory tract, the **mucociliary** (mucus and ciliary) lining of the nasal cavity is the respiratory system's first line of defense. Composed of tiny hairs lining the nose, this defense mechanism filters out the particles inhaled from the environment. The second line of defense is the mucus that lines the turbinate bones (scrolled spongy bones of the nasal passages) in the sinuses and collects particles that get past the nose. These defense mechanisms together trap larger particles from 5 to 10 micrometers in diameter.

As the air we breathe passes through the nose and nasal cavity, it enters the pharynx, where many particles also stick to the mucus on the back of the throat and tonsils. These captured particles can then be eliminated via coughing and sneezing. In addition, the adenoids and tonsils in the back of the throat help trap pathogens for elimination. These lymphoid tissues (tissue from the lymphatic system) also play an important role in developing an **immune system response**, such as the production of **antibodies** to fight off germs.

The lower portion of the respiratory tract also has ciliated cells and mucus-secreting cells that cover it with a layer of mucus. These features work together with the mucus-trapping particles and pathogens, which are then driven upwards by the sweeping ciliary action to the back of the throat where they can be expelled.

Most of the upper respiratory tract surfaces (including the nasal and oral passages, the pharynx, and the trachea) are colonized by a variety of naturally occurring organisms called flora. These organisms (primarily of the staphylococcus group) can help to combat infections and maintain a healthy respiratory system by preventing infectious microorganisms or pathogens from getting a foothold. This phenomenon is known as colonization resistance or inhibition, and occurs because the normal flora compete for space and nutrients in the body. Some flora also produce toxins that are harmful to other pathogenic microorganisms. In rare instances, normal flora can help cause disease if outside factors cause them to become pathogenic or they are introduced into normally sterile sites in the body.

Despite these defense mechanisms, pathogens and particles from 2 to 0.2 micrometers often make their way to the lungs and the alveoli. For example, most bacteria and all viruses are 2 micrometers or smaller. The alveoli, however, also have defense mechanisms to protect against microscopic invaders. In the case of the lungs, these mechanisms are primarily cellular in nature. For example, alveolar **macrophages** are a type of leukocyte that ingest and destroy invading organisms as part of the immune system's response to infection (Figure 10.7). The fluid lining the alveoli contains many components, such as surfactant, phospholipids, and other unidentified agents, that may be important in activating alveolar macrophages. Lymphoid tissue associated with the lungs also plays a role in defending against

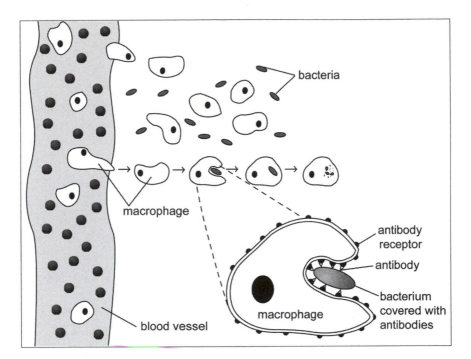

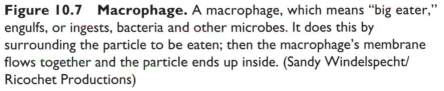

Figure 10.7 Macrophage. A macrophage, which means "big eater," engulfs, or ingests, bacteria and other microbes. It does this by surrounding the particle to be eaten; then the macrophage's membrane flows together and the particle ends up inside. (Sandy Windelspecht/Ricochet Productions)

infections by initiating immune responses. For example, immune system cells, such as B and T cells, represent a local immune response to fight off infections by producing antibodies or activating macrophages.

Disorders and Diseases

While the respiratory system's defense mechanisms are largely effective in battling infections, environmental pathogens can still cause problems when a sufficient "dose" of an infectious agent is inhaled. For example, during cold and flu seasons, a larger quantity of viruses and bacteria are alive and circulating in the air (see Sidebar 10.3). If they enter the

---

SIDEBAR 10.3

Viruses and Bacteria

Because bacteria and viruses cause many familiar diseases, especially in the respiratory system, people often get them confused or think that they are the same type of microbes. In fact, viruses are as different from bacteria as plants are from animals.

Bacteria have a rigid cell wall and a rubbery cell membrane that surround the cytoplasm inside the cell. Within the cytoplasm is all the genetic information that a bacterium needs to grow and to duplicate or reproduce, such as deoxyribonucleic acid (DNA), ribonucleic acid (RNA), and ribosomes. A bacterium also has flagella so that it can move (see Figure 10.8).

Despite the minute size of bacteria, viruses are much smaller. Viruses are surrounded by a spiky layer called the envelope and a protein coat. They also have a core of genetic material, either in the form of DNA or RNA. Unlike bacteria, viruses do not have all the materials needed to reproduce on their own. As a result, they invade cells, either by attaching to a cell and injecting their genes or by being enveloped by the cell. Once inside the cell, they harness the host cell's machinery to reproduce. Viruses eventually multiply and cause the cell to burst, releasing more of the virus to invade other cells.

---

respiratory tract and gain a foothold so that they overcome the body's defense mechanisms and colonize respiratory tract surfaces, the individual will "catch" a cold or the flu (Figures 10.8 and 10.9).

Not all respiratory system problems are caused by infections or environmental assault, such as pollution and cigarette smoke. For example, cystic fibrosis is a genetic disease that can affect the respiratory system by producing an overabundance of thick mucus that can eventually close the respiratory system airways.

Respiratory problems and diseases, some of which are discussed in this chapter, focus on general respiratory ailments that primarily affect the upper respiratory tract. These include epistaxis (bloody nose),

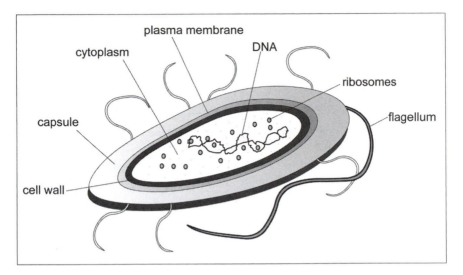

Figure 10.8 Bacterium. (Sandy Windelspecht/Ricochet Productions)

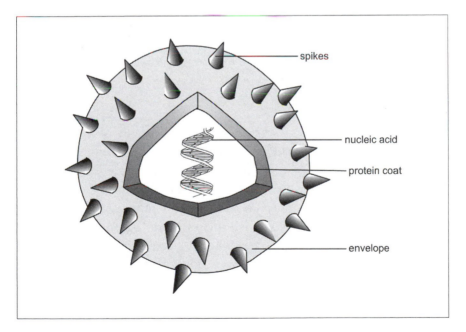

Figure 10.9 Virus. (Sandy Windelspecht/Ricochet Productions)

laryngitis, pharyngitis, rhinitis (also known as hay fever), sinusitis (sinus infection), sleep apnea, strep throat, tonsillitis, and upper respiratory infections (common cold).

Laryngitis

Inflammation of the larynx, or laryngitis, occurs when the vocal cords in the larynx become swollen or scarred. The vocal cords include a central bundle of muscles and various layers of connective tissue covered with mucosa. Any change in these layers can cause hoarseness, low voice, and scratchy throat.

Laryngitis results most often from a viral or bacterial upper respiratory infection. It may also occur when the voice is strained or misused. For example, many people such as auctioneers and singers can develop laryngitis, and overenthusiastic yelling can strain the larynx. Cigarette smoke, dust, or other airborne pollutants are also causes of laryngitis. In rare cases, laryngitis may be linked to a more serious illness such as growths on the vocal chords, allergies, and Horner syndrome, which leads to laryngeal paralysis.

For the most part, laryngitis is temporary in nature, especially if the source is viral, and will heal within a week or so. Most chronic (long-term or frequently recurring) cases are due to continuous voice strain or exposure to environmental pollutants. However, severe cases can result in fever and other respiratory problems.

Rhinitis

Rhinitis is an inflammation of the mucous membrane lining the nose and the sinuses. The primary symptoms include a runny nose, sneezing, and nasal congestion, and may include itchy nose, throat, and eyes. Although rhinitis usually is not a serious problem in terms of impacting a person's overall health, it is an irritating and uncomfortable ailment. However, many cases of allergic rhinitis, particularly when it occurs in children, can lead to various other complications, including chronic otitis media (ear infection), rhinosinusitis, conjunctivitis (eye infection), and sinusitis. Although there are many types of rhinitis, they are broken down into two main categories: allergic rhinitis (more commonly known as hay fever) and nonallergic rhinitis.

Although **allergies** come in many forms, allergic rhinitis is the most common allergic disorder known to medicine. It affects approximately 40 million people in the United States and results in an estimated 10 million lost days of school or work each year. For the most part, it occurs in people who are sensitive to airborne irritants called allergens, including pollen, dust, animal dander, fungus, molds, and grasses. However, allergic rhinitis can also result from allergic reactions to other substances.

In allergic rhinitis, allergens trigger the release of antibodies against the various allergens. These antibodies attach themselves to mast cells, a type of leukocyte or white blood cell that contains various **inflammatory mediators**, including **histamines**, **prostaglandins**, and **leukotrienes**. The mediators cause inflammation and fluid production in the linings of the nasal passages and sinuses.

Although the symptoms of allergic rhinitis are identical to those of the nonallergic form, nonallergic rhinitis does not result from hypersensitivity to allergens. The many types of nonallergic rhinitis include infectious rhinitis, which is caused by an upper respiratory viral or bacterial infection; and eosinophilic rhinitis, which accounts for 20 percent of all rhinitis cases and produces elevated eosinophil (a type of leukocyte) counts in nasal samples, or smears. Scientists have not completely determined its causes.

Sinusitis (Sinus Infection)

Sinusitis, or the common sinus infection, shares a strong resemblance to rhinitis in that it includes inflammation of the membranes that line the nasal passages. However, most sinusitis cases are due to an upper respiratory infection and include inflammation of the paranasal sinuses. In acute cases of sinusitis, the most common causes are viral infections. Fungal infections can cause sinusitis as well, especially in people who have allergies to fungi. In fact, people who suffer from allergies and allergic rhinitis often develop chronic sinusitis. Growths called polyps may block the sinus passages and result in sinusitis in some individuals.

Approximately 37 million Americans each year suffer at least one episode of sinusitis. The most common symptoms in sinusitis are a feeling of throbbing pressure and pain in the sinus areas as well as tenderness in the upper face caused by swelling in the nasal passages. This swelling results in air and mucus getting trapped behind the sinuses, resulting in nasal congestion and blockage and sometimes by a mild fever. A runny nose, headache, and fatigue are also common symptoms.

Sleep Apnea

Apnea comes from the Greek word meaning "without breath." Sleep apnea is characterized by a brief stoppage in breathing during sleep that can last from 10 seconds to over a minute and occurs as many as hundreds of times during the night. First described in 1965, sleep apnea is a common disorder that affects more than 12 million Americans.

The most common type of sleep apnea is obstructive sleep apnea, which occurs when the upper airways become blocked. This blockage usually results from the soft palate at the base of the tongue and the uvula (the small, fleshy tissue that hangs from the center of the back of the throat) collapsing or sagging and partially or completely blocking the airway during sleep. Overweight people often have this type of sleep apnea because of excess tissue in the airway. The other type of sleep apnea is central sleep apnea; it occurs because the brain fails to send signals to the breathing muscles (chest muscles and diaphragm) to make them work.

People with sleep apnea begin breathing again because the brain senses lowered levels of oxygen and increased levels of carbon dioxide in the blood and alerts the body to arousal, that is, to wake up and start breathing again. As a result, people with sleep apnea do not get the same prolonged, restorative sleep as other people. People with sleep apnea will often snore or make choking sounds during the night and repeatedly wake up. This cycle may repeat during the day because the person is often tired from lack of sleep. However, not everybody with sleep apnea snores, especially those with the rare form of central sleep apnea.

Sleep apnea has many serious implications and consequences. One of the most common is increased sleepiness during the day, which hinders concentration and performance in such areas as work and driving a car. Untreated sleep apnea can also cause headaches and serious physical problems, including high blood pressure resulting from the heart pumping harder to make up for oxygen drops in the blood. Eventually, sleep apnea can lead to other cardiovascular diseases and increase risk for heart attack and stroke. A person's memory and sexual functioning may also be affected. Although sleep apnea can occur in anyone, it is most common in men. Overweight people and people who snore often are more likely to develop sleep apnea. Other risk factors include high blood pressure and physical abnormalities in the upper airways. Smoking and alcohol also increase the risk. Because sleep apnea sometimes occurs repeatedly in families, some cases may have a genetic cause.

Upper Respiratory Infections (Common Cold)

According to the National Institute of Allergy and Infectious Diseases, as many as 1 billion upper respiratory infections may occur each year in the United States. More widely known as the common cold, these infections can be caused by more than 200 different viruses. The primary viruses responsible for the common cold are called rhinoviruses (from the Greek word rhin, for nose), which cause an estimated 25 to 35 percent of all colds. (Rhinoviruses may be the main cause of colds because they grow best at 91.4°F or 33°C, which is the temperature of human nasal mucosa.) Other viruses that can cause colds include the myxoviruses (such as the influenza and parainfluenza viruses), coronaviruses, and adenoviruses. Bacterial agents cause approximately 10 percent of colds.

Viruses are transmitted or spread from person to person in several ways. Studies have shown that cold viruses reach their highest concentration in the nasal secretions three to four days after infection, which means this is when the infected person is most contagious and likely to pass on the virus. One common way of catching a cold, or most viral or bacterial infections, is to touch almost anything that an infected person has also touched, sneezed on, or coughed on, from a doorknob to a telephone to their hands. (Some viruses, such as the human immunodeficiency virus, or HIV, cannot be caught in this manner.) After touching the surface, the virus can be transmitted to the body when the person then touches their nose or eyes, which have ducts that drain into the nasal cavity. Inhaling droplets in the air resulting from someone sneezing or coughing close to you is also a common way to catch a cold.

Viruses cause colds when they penetrate the nasal mucosa, after which they enter cells lining the nasal region and the pharynx. Rhinoviruses, for example, bind to a molecule much like a docking system in a space station. Specifically, they contain depressions on their protein shell, sometimes referred to as "canyons," that fit onto surface protein receptors on the nasal cells known as the intercellular adhesion molecules, or ICAMs. This provides the portal for the virus to enter into the cell and begin replicating. It ultimately reproduces thousands of copies of itself, leading to cell disruption and release into the nose, where the infection is further spread to nearby nasal epithelial cells.

Lung Disorders and Diseases

The lungs are unique among internal organs because they are continuously exposed to the external environment. This direct interface with the outside world results in the lungs being assaulted by numerous substances. As a consequence, the lungs are often the most likely organs to be affected by viruses and bacteria, allergy-causing pollen and dust, cigarette smoke, car fumes, and toxic chemicals from factories. Even a naturally occurring substance, like radon gas found in the soil and rocks, can harm the lungs.

Numerous diseases and conditions affect the lungs and impair their ability to provide the body with life-giving oxygen and rid it of carbon dioxide waste. Lung problems and diseases are usually classified according to three major categories, although many lung diseases, such as emphysema, often involve all three. They are:

- Obstructive lung diseases are those diseases in which the airways are narrowed or obstructed, thus decreasing the airflow.

- Restrictive lung diseases occur when the total volume of air that the lungs are able to hold decreases, usually as a result of a lost of elasticity in lung tissue or inability to expand the chest wall during inhalation.

- Diseases that affect the alveoli reduce their ability to diffuse oxygen into the blood.

According to the American Lung Association, lung diseases as a whole are the third-most prevalant killer in America and are responsible for one in seven deaths. Each year, approximately 335,000 Americans die of lung disease. Although lung diseases affect all kinds of people, minority populations, especially African Americans, suffer from a disproportionate share of lung diseases, largely due to increased rates of cigarette smoking.

Ironically, despite the toll that lung diseases take on a person's health and the many deaths caused by them, a large majority of lung diseases could be prevented. For example, the number one cause of lung ailments is smoking cigarettes, a decision that is up to each individual.

Asthma

According to some estimates, some 10–14 million Americans may suffer from asthma, with more than half of the cases occurring in children and teenagers. In fact, asthma is the most common chronic illness in childhood. Asthma is sometimes referred to as "bronchial asthma" because it affects the bronchi, the small air tubes that branch off of the main bronchi and course throughout the lungs. The bronchi are surrounded by bronchial smooth muscle, which contracts or "twitches" as a defense mechanism in reaction to inhaled pollutants, irritants, and other factors. In the case of an asthma attack, these muscles essentially overreact to certain "triggers." The combination of muscle contraction, or spasms, along with bronchial inflammation, swelling, and excess mucus production make the bronchial airways so narrow that the individual finds it hard to breathe, especially to exhale air. Asthma symptoms vary but usually include coughing or wheezing, shortness of breath, and a tightening in the chest.

Asthma is a serious condition and can be a medical emergency in cases of sudden, severe, and prolonged attacks. If the airways become totally blocked, respiratory failure, or suffocation, occurs because the body cannot get enough oxygen.

The triggers for asthma attacks vary from individual to individual. The most common causes are infections (primarily viral) and severe allergies to a wide variety of substances, from pollen and molds to house-dust mites and certain foods. Various irritants, like tobacco smoke and chemical fumes and even cold air or exercise, can cause an asthma attack. In the case of exercise, rapid breathing through the mouth results in air bypassing the nose, which warms air entering the body. As a result, the air reaching the bronchial tubes is cold and triggers an attack in people who are overly sensitive to air temperature. Scientists have also shown that emotional factors like stress can trigger an attack or make it worse.

Chronic Obstructive Pulmonary Disease (COPD)

According to some estimates, approximately 16 million Americans suffer from chronic obstructive pulmonary disease (COPD). The disease, which is characterized by reduced airflow through the respiratory system, is the fourth-leading cause of death in the United States. By far, the primary cause of COPD is smoking tobacco. Other risk factors include exposure to dust and fumes (especially in the workplace such as mines), outdoor air pollution (associated primarily with people who smoke), repeated childhood respiratory tract infections, exposure to secondhand cigarette smoke, and some genetic deficiencies.

Although people with COPD often lead relatively normal lives for many years, COPD is a disease that progressively worsens over time. It is characterized by a chronic cough, spitting or coughing mucus, a loss of breath during exertion or exercise, and a growing inability to exhale air. COPD can encompass many conditions, including chronic asthma; the most common diseases associated with COPD are emphysema and chronic bronchitis.

Emphysema

Like chronic bronchitis, smoking is the overwhelmingly primary cause of emphysema. A deficiency of a protein known as alpha-I antitrypsin (AAT) can also lead to an inherited form of emphysema. Emphysema is the fourth-leading cause of death in the United States and has risen by 40 percent since 1982.

Emphysema does not affect the bronchial tree, but rather causes irreversible damage to the alveoli that cluster in sacs at the ends of the bronchial tree. Among the damages to alveoli are overinflated alveoli, which can fuse with other alveoli to form enlarged alveoli. As a result, the walls between alveoli are reduced in number, and so are the blood vessels that course throughout these walls. This reduction of alveoli walls and surrounding blood vessels results in less surface area to provide for proper gas exchange and oxygenation of the blood. In addition, the surfactant that lines the alveoli within the lungs is damaged, leading to a loss of elasticity so that "stale" air left in the lung is never completely replaced by fresh air. The alveoli can eventually collapse, which causes air to become trapped and results in a greater difficulty in expelling air.

Unfortunately, most people do not pay attention to their symptoms of breathlessness until they lose 50 to 70 percent of their functional lung tissue. In addition to the common symptoms of COPD, other symptoms associated particularly with emphysema include weight loss and an increase in chest size called barrel chest.

Cystic Fibrosis

Cystic fibrosis (CF) is an inherited disease that affects the body's exocrine glands, which produce mucus, tears, sweat, saliva, and digestive juices. Caused by a defective gene, CF changes the chemical composition of these secretions, transforming them from thin and slippery to thick and sticky. Specifically, the gene changes a protein that regulates salt (sodium chloride) movement in and out of cells.

Although CF can affect the liver, pancreas, and the reproductive system, the disease mostly affects the respiratory and digestive systems. In the digestive system, the abnormal mucus can impede the digestive process. However, the disease's effect on the respiratory system is its most dangerous manifestation. The abnormal accumulation of thick mucus in the lungs sets up a breeding ground for bacteria, leading to many respiratory infections. It can also block the airways, and lung disease and respiratory failure are the usual cause of death. People with CF can also develop a collapsed lung, in which air leaks in to the pneumothorax (chest cavity).

Approximately 30,000 people in the United States have CF, which primarily occurs in white people of northern European ancestry. Between 2,500 and 3,200 babies are born with cystic fibrosis each year in the United States. The CF gene is a recessive gene, meaning that two copies of the gene must be inherited, one from each parent. As a result, someone who inherits only one copy of the gene will not develop CF or any symptoms of the disease. However, they do carry the gene and can possibly pass it on, but the disease will manifest itself only if the gene-carrier has a child with someone who also carries the gene. One in 29 people in the United States is a carrier of the CF gene. If two "carriers" have a baby, the child has a 25 percent chance of getting CF and a 50 percent chance of being a carrier. There is also a 25 percent chance that the child will neither get the disease nor become a carrier.

The signs and symptoms of CF may vary, depending on what part of the body is most affected. In infants and young children, the most common signs and symptoms include foul-smelling and greasy stools or bowel movements, weight loss, breathlessness, wheezing, a persistent cough with a thick mucus, and numerous respiratory infections. In infants, older children, and adults, the most common symptom is salty sweat. As a result, a standard test for determining whether someone has CF is the sweat test, which measures the amount of sodium and chloride in a person's sweat. Most people with CF are diagnosed when they are infants or children. Other problems associated with CF include polyps (growths) in the nose, clubbing (enlargement and rounding) of the fingertips and toes, cirrhosis of the liver, and delayed growth.

Influenza ("The Flu")

Influenza, more commonly known as "the flu," is a contagious disease caused by the influenza viruses. Much like the common cold, influenza spreads from person to person in several ways. The primary mode of infection is when a person is near an infected person who coughs and sneezes, or the uninfected person touches something that an infected person has coughed on or touched.

Unlike the common cold, however, the flu often causes severe and even life-threatening illnesses, including bacterial pneumonia. It also can exacerbate other medical conditions, such as asthma, diabetes, and congestive heart failure. Flu **epidemics** and **pandemics** have killed hundreds of thousands of people (see Sidebar 10.4). Even people with the flu who do not suffer serious medical complications are much sicker than the common cold sufferer. In addition to the stuffy nose, sore throat, and dry cough that usually accompany a cold and the flu, people with the flu also suffer from fevers, headaches, body aches, and extreme tiredness.

Each year, approximately 10 to 20 percent of the people in the United States catch the flu. The influenza viruses also kill an average of 36,000 Americans each year and hospitalize another 114,000. Influenza, aided by its major complication of bacterial pneumonia, is the sixth-most common cause of death in the United States.

The influenza viruses are broken down into three major categories: type A, B, and C. The type A viruses are the most common and found in both people and many animals, including birds, pigs, ducks, and horses. Although people can catch the virus from animals, the virus is rarely

SIDEBAR 10.4

The Flu Threat: Now and Then

In the summer of 2009, the World Health Organization (WHO) declared that a global pandemic of the HINI flu (also known as the swine flu) was underway. A pandemic occurs when there is an outbreak of a new influenza A virus; there is little to no immunity to this virus in the human population. This HINI outbreak is in addition to the seasonal flu outbreak that typically affects approximately 10 to 20 percent of the population, according to the Centers for Disease Control and Prevention (CDC). The CDC estimates that there are approximately 36,000 flu-reported deaths reported in the United States every flu season, which is typically late fall to early spring. In addition, there were also confirmed cases of H5N1 flu (also known as bird flu) outside of the United States—primarily in Asia, Africa, the Pacific, Europe, and the Near East.

Flu refers to illnesses caused by a number of different influenza viruses, and can have a wide range of symptoms and effects, from mild to lethal. This latest pandemic does not appear to have the impact of history's other significant influenza pandemic that occurred in 1918. That pandemic, called the "Spanish flu," resulted in an estimated 20 to 50 million deaths. In the United States alone, the Spanish flu epidemic killed more than 500,000 people, including more U.S. soldiers than died in all of World War I. The flu was so pervasive that it reached all areas of the world, including remote northern frozen tundra where it wiped out entire Eskimo villages.

The Spanish flu had the ability to kill young and healthy adults, who in a matter of hours could develop fevers of 105°F and become so weak they could not walk. As the virus swept throughout the United States, fear became pervasive. Doctors were shocked when they performed autopsies on those who died and found a bloody and foamy liquid that entered the lungs in such quantities that it caused people to drown, or suffocate, in the mucus-like fluid. Another flu epidemic comparable to the Spanish flu has not occurred since 1918. If such an epidemic occurred today, an estimated 1.5 million Americans would die.

In addition to the 2009 and 1918 pandemics, other pandemics have occurred. The "Asian flu" pandemic of 1957–1958 caused 70,000 deaths in the United States, and the "Hong Kong flu" killed approximately 34,000 Americans in 1968–1969. Potential pandemics that never developed include the "Swine flu" outbreak of 1976 and the "avian flu" outbreak of 1997, in which 19 people in Hong Kong came down with a type of influenza infection that was thought to occur only in birds.

The potential for a new flu virus to emerge and result in a deadly epidemic that can quickly become a pandemic is very real, especially because of the growing number of people who travel worldwide in a matter of hours. Scientists are constantly researching influenza viruses in an effort to develop better vaccines. In the United States, the CDC and local public health agencies also maintain surveillance systems that monitor such factors as weekly pneumonia and influenza deaths and overall influenza activity.

---

spread this way. Type A viruses are the causes of the most serious epidemics and pandemics. The influenza B virus is also very contagious and causes large epidemics, but the disease and its complications are much milder than those caused by type A viruses. The influenza C viruses are much more mild than either A or B viruses and are not believed to cause epidemics.

By far, the best approach is to avoid the flu by getting a flu vaccination. Because flu viruses change over time, different vaccines are developed each year based on the viruses circulating at the time. Vaccines are made from killed, or deactivated, flu viruses grown in chicken eggs. However, even if a person receives a flu shot, they can still catch the flu because new strains of the virus may appear during the course of the flu season, which is when most flu cases occur because, it is believed, people are indoors more where they are in closer contact with other infected individuals and breathing recirculated air. Nevertheless, the flu vaccine usually leads to milder symptoms and complications even when it encounters a new viral strain. The flu vaccine can have side effects. Even though most of them are mild, certain people should consult their doctors and avoid vaccination, especially if they are allergic to chicken eggs, have previously had a serious reaction to a flu shot, have a paralytic disorder such as Guillain-Barré Syndrome, or are currently sick with a fever.

Lung Cancer

Although lung cancer was a rare and virtually unknown disease in the United States prior to 1900, the incidence of lung cancer has been steadily rising in correlation with the growing popularity of cigarette smoking since the 1930s. Not only is lung cancer one of the most common cancers in the United States, it is the leading cause of death due to cancer in American men and women. In 2002, about 150,000 people died from lung cancer, and approximately 170,000 new cases of lung cancer are diagnosed each year.

Lung cancer occurs when cells in the lungs begin to divide and grow abnormally. Once this process gets out of control, abnormal tissues called tumors are formed. Tumors can be either benign, meaning that they do not spread and are noncancerous, or malignant, meaning that they are cancerous and will continue to spread, often throughout the body.

Although many types of lung cancer occur, they are grouped into two primary categories called non–small cell lung cancer and small cell lung cancer. This division is based on how the cancer looks under a microscope,

---

SIDEBAR 10.5

Did You Know? Cigarette Smoking

Cigarette smoking is the single most preventable cause of premature death in the United States and accounts for one out of every five deaths, killing more than 430,000 people in the United States each year. Compared to people who do not smoke, men who smoke increase their risk of dying from lung cancer by more than 22 times and from chronic bronchitis and emphysema by nearly 10 times. Women, in whom lung cancer increased dramatically between 1960 and 1990, have a 12-times-increased risk of dying from lung cancer and a 10-times-increased risk of dying from chronic bronchitis and emphysema. Exposure to secondhand, or environmental, tobacco smoke also causes an estimated 3,000 deaths from lung cancer in American adults who do not smoke. Maternal smoking has also been strongly associated with adverse respiratory effects in children, and there is evidence that it even affects the child while it is in the womb.

and not because of the tumor size. Non–small cell cancer is the most common form of lung cancer and includes squamous cell lung cancer, adenocarcinoma, and large cell carcinoma. These cancers usually spread more slowly than small cell lung cancer, which is also more likely to spread to other parts of the body.

Smoking cigarettes causes most cases of lung cancer (see Sidebar 10.5). Cigar smoking also causes lung cancer, but not as often because people usually smoke fewer cigars and do not inhale cigar smoke as deeply. More than 90 percent of all lung cancer deaths are related to smoking tobacco and inhaling the more than 4,000 chemicals that tobacco contains, many of which are carcinogens. In addition, many lung cancer deaths are due to people being exposed to secondhand environmental smoke—that is, the smoke blown into the air by smokers. Not everyone who smokes gets lung cancer, and some people develop the disease because of other reasons, including exposure to radon, asbestos, and certain air pollutants and other substances. Lung diseases such as tuberculosis also increase the risk for developing lung cancer.

Summary

The respiratory system's complex system of organs and tissues in the upper and lower respiratory tracts regulate the respiration process. The organs of the upper respiratory tract are the nose and nasal cavity or passage, the pharynx or throat, and the larynx or voice box. Located in the lower respiratory tract are the trachea or windpipe, the bronchi, the alveoli, and the lungs.

The breathing process starts when nerve impulses stimulate the breathing process, prompting air to move into and out of the lungs through nasal passage, down through the throat, and into the lungs. This air is filled with oxygen, and when it reaches the lungs, gas exchange occurs—oxygen is exchanged for carbon dioxide. Known as external respiration, this exchange occurs between the lungs and the blood. Now the blood is rich with oxygen, and the circulatory system transports this blood throughout the body's tissues, where the exchange process repeats itself. This time, however, the exchange occurs between the blood and the cells, with the blood passing oxygen into the cells and carrying away carbon dioxide, which will be pushed out of the body through the lungs. This is called internal respiration. The third type of respiration is called cellular respiration, and is when oxygen is exchanged throughout the cells to perform varies functions.

One of the distinct characteristics of the respiratory system is that it is in constant contact with the outside environment, particularly the air. Because of this, it is exposed to substances such as bacteria, virus, and chemical pollutants. Therefore, a significant number of diseases and disorders are associated with this system of the body, including influenza, asthma, and even lung cancer.

11

The Skeletal System

Evelyn Kelly

Interesting Facts

- Eighty bones protect the vital organs of heart, lungs, spinal cord, and brain.

- Children with broken bones heal much faster than adults. A bone that requires three to five months for healing in an adult will mend in four to six weeks in a child.

- The spinal column consists of a series of 26 individual bones, or vertebrae.

- Motorcycle accidents account for one injury to the skeletal and muscular systems in every 7,000 hours of biking; horseback-riding accidents account for one injury in every 2,000 hours of riding—three and one-half times more than motorcycling.

- About 6.8 million people seek medical attention each year for injuries involving the skeletal system.

- Throughout the day, the discs in the spine are squashed, making people shorter when they go to bed than when they wake up.

- Osteoclasts consume old and worn bone matter; osteoblasts manufacture new bone tissue. Both are important to good bone health.

- Sports medicine has been around since ancient times. To stay alive to fight, warriors had to keep in top physical condition.

- Nearly 40 million Americans—or 1 in 7—have arthritis, including 285,000 children.

- The average person will walk about 115,000 miles during a lifetime; that accounts for more than four jaunts around the equator on the feet.

Chapter Highlights

- Skeleton system's primary functions

- Different types of bones

- How bones move

- Protective functions of bones

- The axial system

- The appendicular system

- How joints, ligaments, tendons, and cartilage are vital to the skeletal system

Words to Watch For

Abduction	Cancellous bone	Epiphysis
Adduction	Collagen	Equilibrium
Adipocytes	Compression	Extension
Amphiarthroses	Coronal plane	Flexion
Anterior	Cortical bone	Fossa
Appendicular system	Cuboid bones	Hematopoiesis
Arthritis	Diaphysis	Intracapsular
Axial system	Diarthroses	ligaments
Bipedal	Dislocation	Joints
Bursa	Distal	Lateral

Ligaments	Plantar	Symphysis
Medial	Posterior	Synarthroses
Midsaggital plane	Proximal	Synolvial fluid
Opposable thumb	Rotation	Trabeculae
Osteology	Sagittal plane	Trabecular bone
Palmar	Sesamoid bone	Transverse plane
Pituitary gland	Superior	

Introduction

From ancient times, bones have fascinated human beings. In fact, bones are an integral part of ancient documents, literature, and art. The Hebrew Old Testament tells how God caused a deep sleep to come over Adam so that He could take one of Adam's ribs to make woman. Adam recognized, "This is now bone of my bone" (Genesis 1:28).

In the second century, the Greco-Roman physician Galen (129–216?) told medical students in his book *On Bones* the importance of studying osteology. However, Galen was able to study only the bones of animals and criminals who had not been buried. Human dissection was not permitted at that time.

Bones have often appeared as a symbol of danger or death. In the 1500s, pirates who plundered and terrorized ships and coastal towns hoisted flags bearing a skull and crossbones. The same symbol was at one time printed on labels of poisonous materials or used to mark hazardous places.

Throughout history, bones were associated with death because bone was considered to be dead. In fact, the word skeleton comes from the Greek word skeletos, meaning "dried up." It was not until the awakening of scientific investigation in the eighteenth century that an English surgeon, John Hunter (1728–1793), discovered that bone is living, dynamic, and changing.

This chapter will explore this living and dynamic system, from its function to its anatomy.

Functions of the Skeletal System

A formless jellyfish floats on the water, wafting back and forth with the wind and waves. Without bones, human beings would be just like this creature—gigantic, massive blobs that would move only by inching along the ground. Without bones, we would not be able to stand up straight, run, do handsprings, or even hear.

The skeleton provides shape and support to all the other body systems. In addition, it allows movement, protects body **tissues** and organs, stores important materials, produces valuable blood cells, and holds a record of our past development, diet, illnesses, and injuries.

Providing Shape and Support

In the human body, 206 bones are intricately arranged to keep the body upright. The skeleton is both rigid and flexible, enabling internal organs to defy the forces of gravity. The unique architectural plan makes the scaffold on which other body parts are hung and supported. One could even say the human being is a model of architecture and design. In fact, Marcus Vitruvius (first century BCE), an ancient Roman architect and engineer, advised his students who were designing symmetrical temples to study the human body because when a person's arms and legs are extended, he or she can touch a square with four corners and form a perfectly circular arc.

Body Plan

Compared with the nearest related primate, the gorilla, Homo sapiens is completely erect. Although the skull is at one end of the body, it functions as the center and main point of reference. As with any well-designed structure, form follows function. The classical anatomical form of Homo sapiens is standing erect, head straight, facing the observer, with arms at the side and palms facing forward. A midline perpendicular to the ground divides the body into left and right halves. This line represents the **sagittal plane**. A **transverse plane** runs parallel to the ground or floor.

To understand and read anatomy, one must be familiar with words that describe the positions of the body plan (Figure 11.1). The following terms

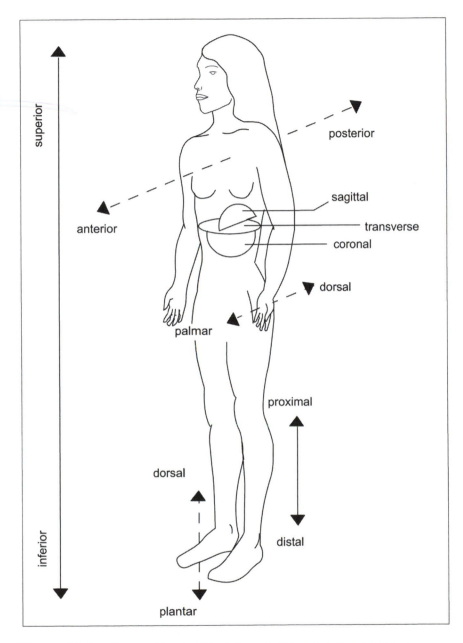

Figure 11.1 Planes. (Sandy Windelspecht/Ricochet Productions)

define the relationship of shape and support that create the framework of Homo sapiens:

- **Superior** indicates toward the head; **inferior** indicates away from the head.

- **Anterior** refers to the front part of the body or body part; **posterior** refers to the back of the body or body part.

- **Medial** indicates toward the midline; **lateral** indicates to the side away from the middle.

- **Proximal** indicates closer to the torso; **distal** indicates farther away from the torso. For example, the hand is distal to the forearm because it is farther away from the torso than the arm.

- **Palmar** indicates the palm of the hand; dorsal indicates the back of the hand.

- **Plantar** indicates the sole of the foot.

- **Coronal** plane divides the body into front and back portions.

The entire skeletal structure works together to manage the forces created by the upright position. When prehistoric animals walked on all fours, their back legs formed a right angle with the spine. The dog and cat reflect this general design. In the chimpanzee, which walks in a semi-upright position, the legs form less than a right angle. The upright position of the human causes the legs to form a straight line that runs through the backbone, the pelvis, and down each leg. When walking, as a leg on one side swings back, the arm on the same side swings forward. When taking a step, the opposite happens. The alternating movement keeps the body weight right in the center.

When a person uses two feet, the stresses of walking, running, and jumping are calculated with precision. For example, when a person stands up, the hip takes half of the body weight, and the pull of stabilizing muscles can multiply this weight six times. When a person runs or walks, each hip alternates carrying the full body weight. Bones are extra strong at joints where large **compression** forces are generated.

The cranium and vertebrae support the brain and spinal cord. Whereas the cranium, or skull, shapes and supports delicate brain structures, the spine is the literal "backbone" of the human body. It forms a supporting rod for the head, arms, and legs. In other mammals, the spine is a horizontal girder taking the weight off the chest and abdomen of animals that move on all fours. When the human spine is viewed from the side, an S-shaped curve is visible; this curve aids in balancing parts of the body over the legs and feet when standing. The chain-link arrangement of the vertebrae allows only a small amount of movement for each link; but when all the individual movements are added up, the spine is capable of making large, complex shapes and contortions. The S-shape helps bring the centers of gravity of the head, arms, chest, and abdomen above the legs. Thus, the body as a whole is well balanced.

A Framework of Different Shapes

Each bone is designed to support, protect, and shape the body. For example, relative to body size, the human skull protects one of the biggest brains in the animal world. The features of different bones allow for efficient form and function. Bones may be classified by shape.

Long Bones

Like tubular furniture, the tubular structure of long bones makes them both strong and light. The outer shell of bone is made of compact **cortical bone**. The spongy center has little **trabeculae**, or beams, that act as strengthening girders of **cancellous bone**. The inner casing of **trabecular bone** is arranged along lines of force with calcified fibers as they transmit the force into **tendons**, the fibrous bands that join muscles to bones. Bones are thickest in the middle to support areas where forces are strongest. Weight for weight, bone structure is stronger than that of a solid rod. The lower extremities of bones are longer and stronger than the upper extremities. The purpose of the long bone of each leg, the femur, is to provide form and support and create an interconnected set of levers and linkages that allow movement.

Short Bones
Found in the wrist and ankle, these spongy, **cuboid bones** (shaped like cubes) are covered with a thin layer of compact bone. This arrangement permits shock absorption, movement, elasticity, and flexibility.

Flat Bones
These bones in the ribs, the crest of the hip, the breastbone, and the shoulder blade are sandwiches of spongy bone between two layers of compact bone. They protect and provide attachment sites for muscles.

Irregular Bones
Bones of the skull, face, vertebrae, and pelvis, as well as others that do not fit into another category, are referred to as irregular. Usually consisting of spongy bone with a thin compact bone exterior, they support weight and dissipate loads.

Sesamoid Bones
These are short bones embedded within a tendon or joint capsule. An example is the patella (kneecap), which allows the angle of insertion of a muscle to be altered.

The type of support provided by the bones reflects gender differences. Although men and women have the same number of bones, women's skeletons are lighter and smaller. Women's shoulders are narrow whereas their hips are broad and boat-shaped to accommodate a growing fetus. In men, proportions are reversed: broad shoulders and slim hips.

Allowing Movement
The skeletal system is an engineer's dream of a moving machine. "Mechanical science is of all the noblest and most useful, seeing that by means of this all animate bodies which have movement perform all these actions," said Leonardo da Vinci (1452–1519). Combining art and anatomy, he was one of the first to study how the interaction of the skeletal and muscular systems allows bodily movement.

Many parts of the skeletal system help the body to move. Tendons connect muscle to bone; **ligaments** connect bone to bone. Bones meet each other at **joints**. Muscles cause movement at joints by working in pairs.

When one muscle contracts, the other extends. For example, the contraction of the biceps muscle in the front of the upper limb (humerus) and the relaxation of the triceps muscle behind this bone cause the elbow to bend. When the triceps contracts and the biceps relaxes, the arm straightens.

Rigidity of the bones allows for movement. The attached muscles at the joint permit freedom of movement in a variety of planes and in almost any direction.

Scientists are slowly learning more about how the human machine works. The field of bioengineering combines engineering principles with the anatomy of the living body to understand how movement occurs. For example, in a procedure called gait analysis, subjects are fitted with diodes, or electronic sensors, that send out pulses of infrared rays of light as the subjects walk a prescribed route in the laboratory. Cameras sensitive to the emitted light follow the walkers' trails and record the positions relative to a fixed background. The cameras shoot as many as 315 frames per second. When this information is analyzed on a computer, a three-dimensional picture reveals if the person has problems with walking that arise from a skeletal deformity.

Bones as Levers

The science of physics is very important to human anatomy. Each time a person moves, the laws of physics come into play. The mechanics of movement integrates the laws of physics into biology. The skeleton is first and foremost a mechanical organ because one of its primary functions is to transmit forces from one part of the body to another. The tissues must bear loads without being damaged, and the skeleton must withstand very high forces because muscles can contract only a small percentage of their length. With the help of muscles, the bones pull and push, creating the action of a lever (Figure 11.2). The term fulcrum refers to the point of support at which a lever turns in raising or moving something. This is the pivot point of the lever.

The underlying principle of physics that every machine reflects is the conservation of energy. Thus, to simplify movements, levers save rather than spend forces. Moreover, the lever is one of the simplest machines. At the same time work (force) is exerted on one end of the lever, the other

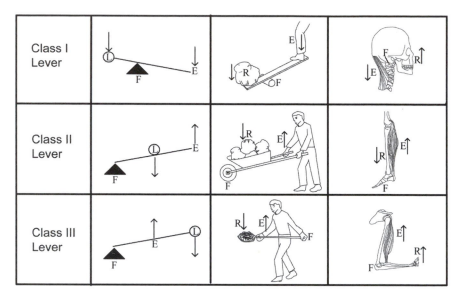

Figure 11.2   Levers. (Sandy Windelspecht/Ricochet Productions)

end moves the load. In the human body, muscles and bones function as one of three types of levers.

Class 1 Lever
In this class, the fulcrum is between the force and the load. A seesaw on a playground is an example. When force is exerted at one end, the load (the child) at the other end is lifted. In the human body, an example of a class 1 lever is the arm bending at the elbow. The elbow is the fulcrum. Contraction of the biceps muscle exerts force; the mass of the forearm bone is the resistance force. Another example is the head raising to look up. Muscles in back of the neck contract; bones of the face resist.

Class 2 Lever
In this class, the load is between the effort force and the fulcrum. For example, exerting force on a long steel bar placed under an automobile frame lifts the car. Raising the foot to stand on tiptoe is an illustration in the human body. Gliding joints at the ankles, or tarsals, provide the fulcrum.

Pushing down on the shin bone, or tibia, and the other leg bone, the fibula, provides resistance, while contracting the soleus muscle in the calf of the leg provides the effort force.

Class 3 Lever

In this class, the fulcrum is at one end and the load is at the other end, with the effort force applied between. In the human body, the fulcrum is the elbow and the load is the hand. When the biceps move the hand, they function as a class 3 lever. This type of lever increases distance at the expense of force. When the biceps muscle moves a short distance, the hand moves a greater distance. The input and output forces are on the same side of the fulcrum and have the same direction.

Motion through Levers

How all these parts work together is demonstrated by a biomechanical marvel, the knee. The knee bone, or patella, hangs free from the center of the body and bends, glides, and rotates with the stresses that motion puts on it. Although 4 major ligaments and 13 muscles support it, the knee is the most vulnerable of all joints because outside forces may displace these structures.

When a person takes a step, the ligaments and muscles contract, relax, twist, and turn. Two long bones, the thigh and shin, sit between a piece of shock-resistant cartilage called meniscus. The crescent-shaped meniscus cushions and absorbs the shock of movement when walking or running.

To understand this function of movement, one must know several terms related to the action of bones and joints:

- **Flexion** involves bending or decreasing the angle at the joints. When the calf bends back toward the thigh, flexion occurs. When a bodybuilder flexes his muscles, he changes the angle of his bones at the joints.

- **Extension** is the opposite of flexion, with bones straightened to a 180-degree angle—a straight line.

- **Rotation** involves turning a body part on an axis. Just as the earth turns, the entire body may turn. However, a single body part (such as an arm or leg) cannot turn in a complete circle of 360 degrees because doing so would tear tissues such as blood vessels and nerves.

- **Abduction** involves drawing away from the midline of the body. Lifting up the arm at the shoulder joint moves it away from the body.

- **Adduction** involves moving toward the midline or trunk. Dropping the arm at the shoulder joint moves it back toward the body.

Protecting Tissues and Organs

The skeletal system protects the tissues and organs from the hard knocks of life. Although tremendous forces are constantly bombarding human bodies, bone is one of the strongest materials that nature has devised. One cubic inch can withstand loads of up to 20,000 pounds, which is about four times the strength of concrete.

The cranium, or skull, protects the soft, delicate brain, which has the consistency of custard. The strong cranial case, which has an internal volume of about 2.5 pints, contains the organ that acts as "chief executive officer" of the body by controlling information from the outside world and responding to this information. Also protected are the organs of sight, hearing, smell, and taste. Eyes are set in sockets in the skull to stabilize the delicate structures. The delicate middle and inner ear not only are surrounded by bone, but have three tiny bones that transmit sound. Through the bony recesses of the nose, air with vital oxygen passes on the way to the lungs. The jaws and teeth crush nourishing food on the way to the digestive system.

Attached to the cranium is the vertebral column, which consists of a series of small bones, stacked on top of each other like a tower of spools. Large holes in the vertebrae line up to form a bony tunnel or canal. Nerves enter and leave the tunnel through gaps between neighboring vertebrae. Together, the vertebrae protect the delicate spinal cord, the important message cable between the brain and the other body parts. In front of the spinal cord, the rib cage and breastbone protect the heart, lungs, and part of the digestive system.

Producing Blood Cells

When a person gets a cut, a red fluid comes through the wound in the skin that contains red blood cells manufactured in the bones. Whenever there is a risk of infection or even a common cold, the body's defenders—white blood cells—are also made by the bones.

Indeed, an important function of the skeletal system is the production of blood cells. Marrow is a Latin word that means "middle"; hence, **bone marrow** is located in the middle, spongy part of bone. There are two types of marrow: Red marrow produces red blood cells, or **erythrocytes** and white blood cells or **leukocytes**; yellow marrow produces fat cells, or **adipoctyes**.

Red bone marrow is the site of **hematopoiesis**, or blood cell formation. In the adult human, this production takes place in bones such as the vertebrae, sternum, ribs, and pelvis and also at the ends of the upper arm (humerus) and the upper leg (femur). Red cells in the marrow make a substance called heme, which is an iron-containing nonprotein portion of **hemoglobin**, the red part of the erythrocyte. Hemoglobin contains iron and carries oxygen from the lungs to tissues. About 175 billion red cells per day are made and released according to the demands of the body. The red marrow also produces white blood cells. About 70 billion per day of these important cells are needed for the body's defense. Also, about 175 million platelets per day are produced. These blood cells are important for clotting. According to demand, the system may increase production five- to tenfold.

In newborns, red bone marrow fills in most marrow cavities. In older adults, much of the red marrow has been converted to yellow marrow. Long bones are filled with yellow marrow, which is mostly fat. In certain conditions like **anemia**, which occurs when there is a shortage of oxygen-carrying red blood cells, yellow marrow can be converted to red marrow for the manufacture of more red blood cells.

Storing Materials

The bones are keepers of the minerals calcium and phosphorus. Deposited minerals account for about 50 percent of a bone's volume and 75 percent of its weight. In fact, 97 percent of the body's calcium is stored in bone.

Not only does the skeletal system store minerals, but it acts as a reservoir that maintains the **equilibrium**, or balance, of calcium and phosphorus in the bloodstream. Calcium **homeostasis** is very important for bodily functions. For example, too many calcium ions, or charged particles of calcium atoms, in the bloodstream can cause heart attacks; too few can cause respiratory problems. In the bones, these substances are available for rapid turnover when needed. For example, in pregnancy, a growing fetus has a high demand for calcium. Storage in the bones makes available an extra supply. Likewise, after menopause, when menstrual activity ceases, changes in hormones may impair a woman's calcium and phosphorus levels, causing the minerals to leach out and leaving brittle, osteoporotic bones.

Many hormones play a role in this storage system. These include estrogen, testosterone, thyroid hormone, adrenal gland hormones, insulin, and growth hormone.

Giving Clues to the Past

The skeleton's durability is staggering. Even after death, it has many uses. Archaeologists, anthropologists, and forensic scientists can glean valuable information from bones as to what happened in the past. For example, fossils (bones that have turned to stone) offer a broad outline of how the human face evolved. Fossils of Australopithecus, the Southern Ape that existed 2 to 3 million years ago, can be compared to those of Homo sapiens neanderthalensis, or Neanderthal Man, of 100,000 years ago. Both of these can be compared to the first Homo sapiens, or Thinking Man, who lived 40,000 years ago. Gradually, the face became flatter, the teeth smaller, the chin less protruding, and the forehead more domed to house a larger brain.

Bones also tell the story of humans' adaptive mechanisms to the environment over time. Bones often survive the process of decay and are important in the following ways:

- They provide evidence of fossil man.

- They form the basis of racial classification in prehistory.

- They give information about culture and people's worldview.

- They are major sources of information about ancient disease and causes of death.

- They help solve forensic crimes by providing evidence to detectives.

The human skeletal system is a marvel of mechanical and architectural design. It is intricately structured to keep the body straight and upright, but flexible enough to permit great freedom of movement. It is strong enough to protect vital organs while also producing blood cells and storing and regulating minerals. As a source of data, bones have made possible the study of the evolution, history, and culture of Homo sapiens.

Bones of the Central Skeleton: The Axial System

Osteology—the study of bones—is important to many scientific disciplines, including archaeology, physical anthropology, geology, paleontology, anatomy, medicine, and forensics. In fact, knowledge of the framework of the skeleton is essential not only for scientists, but also for lay people. How the body's framework functions is essential to developing good health habits and care of the skeletal system.

The sturdy scaffold of the human body is made up of 206 bones. Softer tissues and organs are attached to this structure. Actually, the skeleton, directly or indirectly, supports or connects to all body parts. Bones are grouped into two categories: the **axial skeleton** and the **appendicular skeleton**. This section discusses the axial skeleton; bones of the skull, vertebral column, ribs, and breastbone.

Special Bone Structures

Some structures are important in understanding both the axial and the appendicular skeletons. These structures include sutures and processes.

Sutures

During birth, a baby's head is squeezed as it passes through the birth canal. In order for the baby to be born safely, its skull is not solid bone. A **membrane** covers the areas where the skull bones have not yet grown together. The structures known as fontanels, or soft spots, allow the bones of the skull to mold, slide, and overlap to minimize danger to the delicate brain during birth. These soft areas disappear completely by age 2, closed by structures called sutures.

Sutures could be referred to as bone zippers as they occur where the bones of the skull come together along serrated and interlocking joints. The areas appear as irregular gaps before the age of 17 but grow together as the person gets older. By age 30 or 40, the sutures gradually fade away. Looking at these sutures is one way of telling the approximate age of a skull (Figure 11.3).

As the sutures turn into bone, the process (the outgrowth of bone) begins inside the skull and then knits together toward the outside. Most sutures are named from the bones that grow together to form them. For example, the ethmoidofrontal suture joins the ethmoid and frontal bones. Other special names for sutures follow:

Coronal: Between the frontal and parietal bones

Sagittal: Between the two parietal bones

Basilar: Between the occipital and the sphenoid bone

Squamosal: Between temporal and parietal bones

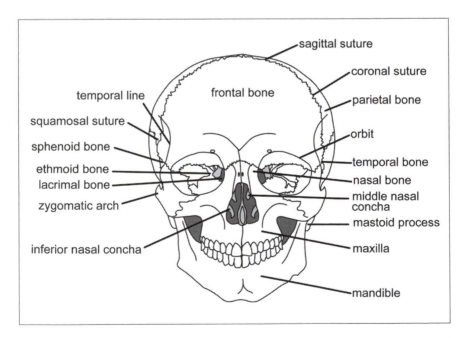

Figure 11.3   Skull, anterior. (Sandy Windelspecht/Ricochet Productions)

Processes

A process is a projection or outgrowth of bone or tissue. On any given bone, depressions and holes may also be present. These structures provide for the attachment of muscles, help form joints, and act as passageways for blood vessels and nerves.

Bone processes include the following structures:

- *Head*—a large round part that joins with another bone; the term articulation refers to the joining of one bone to another

- *Shaft*—the principal part of a long bone

- *Neck*—the narrow part of a bone between the head and shaft

- *Spine*—a sharp slender process, as seen on the back of the shoulder blade (scapula)

- *Condyle*—a rounded, knuckle-like process located where one bone articulates with another

- *Crest*—a very narrow ridge of bone

- *Trochanter*—a large projection for the attachment of muscles

- *Depression, or fossa*—a shallow hole in the surface of the bone

The Axial Skeleton

The axial skeleton—the central supporting portion of the body—is composed of the skull, vertebral column, ribs, and breastbone. The term axial is derived from the word axis, a real or imaginary straight line that runs through the center of a body. Axis may also refer to a structure around which other objects rotate. The axial skeleton has 80 bones.

The Skull

The skull is divided into the cranium and facial bones. Oval in shape and wider behind than in front, the skull rests on the top of the vertebral column. The human skull starts life as a "jigsaw puzzle" of about 30 pieces held in cartilage and membranes. During embryonic development, these pieces gradually grow together to form a solid case. There are six fontanels at birth; the pulsing of the baby's blood system can be seen in the upper-most fontanel. The cranium, which lodges and protects the brain, consists of eight bones. The names of these eight bones, along with their common names or locations, follow:

- *Frontal*—the forehead

- *Occipital*—lower back of the skull

- *Sphenoid*—large bone between the occipital and ethmoid in front and temporal bones at the side

- *Ethmoid*—inner part of the eye socket and back of the nose

- Two sets of *parietal*—top and side of the skull

- Two sets of *temporal*—temple areas on the side of the skull

These flattened or irregular bones do not move—with one exception, the mandible (jaw). They are joined at points called sutures.

Frontal Bone   The frontal bone forms the forehead and the upper portion called the squama in the forehead region, and a horizontal portion that forms the roofs of the orbital and nasal cavities (Figure 11.4)

The outer surface of the squama is convex and usually shows the remains of the frontal, or metopic, suture that divides the bone in two. On each side of this suture, located about 3 centimeters above the eye socket, is a rounded elevated area called the frontal eminence. These protrusions vary in size and are usually larger in men than in women. Sometimes they are referred to as supraorbital ridges. The frontal bone articulates with 12 other bones.

Occipital Bone   The occipital bone forms the posterior, or back, surface of the skull. It is shaped like a trapezoid—a four-sided figure with two parallel sides and two nonparallel sides. The structure seems to curve in on itself. A large oval hole, or aperture, called the foramen magnum pierces the bone. This hole, the largest foramen, allows the spinal cord to pass through the bone.

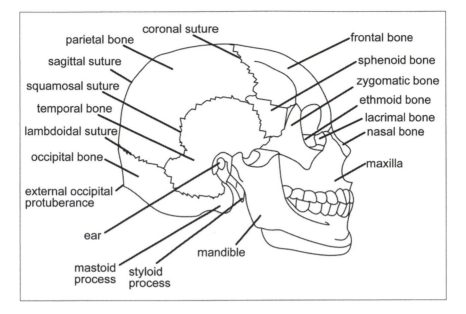

Figure 11.4 Skull, lateral. (Sandy Windelspecht/Ricochet Productions)

The external surface of the occipital bone curves out in a convex area, producing a protrusion on either side to which ligaments are attached. The lateral, or side, parts of the occipital bone rest at the side of the foramen magnum (Figure 11.5).

Sphenoid Bone One of the most difficult bones to describe is the sphenoid. A number of features and projections enable it to be viewed from various points. A single bone, it runs through the **midsagittal plane** and connects the cranium to the facial bones.

The sphenoid bone is a hollow body that contains the sphenoid sinus and three pairs of projections. On the inside of the sphenoid is a small, saddle-shaped shelf where the **pituitary gland** rests. One section of the sphenoid bone called the smaller lesser wings has a hole that allows the optic, or second cranial, nerve to pass through. Other processes run along the back portion of the nasal passages toward the palate, or roof of the mouth. Muscles run from these attachments to the internal, or medial, surface of the mandible, or jawbone. These muscles provide the grinding motion of chewing.

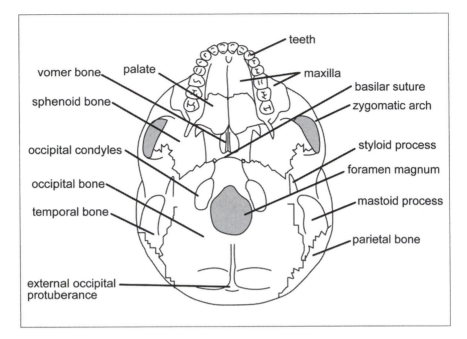

Figure 11.5 Skull, interior. (Sandy Windelspecht/Ricochet Productions)

Ethmoid Bone   The ethmoid bone is a light and somewhat spongy bone that is cubical. A single bone, it runs through the midsagittal plane that connects the cranium to the facial skeleton. Unlike the sphenoid, one cannot see it from various views.

The bone has paired projections called the Crista Galli, or Cock's Comb. Several plates make up this single bone, which cradles the nerve of smell, separates the nasal passage, and forms part of the eye socket.

Ethmoid Notch   The ethmoid notch separates two orbital plates. In front of the notch are the openings to the frontal air sinuses. A sinus is a cavity within a bone or organ. In the interior of the cranium, these cavities are vein channels that move blood from the brain. External sinuses—such as those found in the frontal, sphenoidal, ethmoid, and maxillary (upper jaw) areas—are hollow spaces in the bone. They are connected to the nasal cavities that contain air.

Parietal Bones Between the frontal and occipital bones are two parietal bones. As the two bones unite, they form the sides and roof of the cranium. Each bone is roughly quadrilateral and has two surfaces, four borders, and four angles.

Temporal Bones The temporal bones are paired cranial bones situated at the side and base of the skull. They are located below the parietal bones and form part of the sides of the base of the cranium. Each temporal bone contains the middle and upper portion of the hearing mechanism. One area of the temporal bone forms the mastoid process, a conical projection. The mastoid can be felt by placing the hand behind the ear. The mastoid process is larger in men than in women. Part of the temporal bone extends out to form the zygomatic arch (see the next section) of the facial bones.

Facial Bones

All bones of the face are paired except the vomer, mandible, and hyoid. The paired bones are the following:

* Zygomatic

* Maxillae

* Nasal

* Lacrimal

* Palatines

Zygomatic The cheekbone is made up of two bones; the zygomatic, and a fingerlike projection from the temporal bone. This joining is called the zygomatic arch. Also called malar or jugal, each zygomatic bone articulates (joins) with surrounding bones, one on each side.

Maxillae The maxillae, or upper jawbone, are paired facial bones that join to form the hard palate in the roof of the mouth. They also contain the upper teeth.

Nasal Some facial features are composed of both bone and cartilage. The nasal bones are small rectangular bones that form the upper part of the bridge of the nose. Cartilage forms the lower part of the nasal frame. Cartilage deteriorates after death. This is why one never sees a skeleton with a nose.

Lacrimal The lacrimal bones are located behind and lateral to the nasal bones. These small and fragile bones help form the eye orbit and part of the nasal passage. They also contain the fossae, or holes, housing the lacrimal duct that connects the medial corner of the eye to the nasal passage. This duct enables tears from the eye to enter the nasal passage.

Ossicles The human body's tiniest bones are in the ear. Three little bones called ossicles are located in each middle ear. The bones are named for their appearance:

- The malleus, or hammer, is about 0.32 inch long.

- The incus, or anvil, looks like a small version of the metal table used by blacksmiths to hammer iron tools.

- The stapes, or stirrups, appear as a saddle about 1.2 inches long.

The mallet-shaped handle of the malleus attaches to the inner surface of the eardrum. The head of the hammer fits into a tiny socket at the base of the anvil. Small ligaments between these two bones bind them firmly together. A long process of the anvil joins with the head of the stirrup. Hearing occurs when sound waves strike the eardrum and set the ossicles in motion. Moving through the bones, the waves result in a rocking motion of the stirrups that oscillates against the membrane covering the opening (called the oval window) in the inner wall of the middle ear. At birth, the ossicles are completely developed. They do not change in size as a person grows.

Vomer The vomer is a thin, flat bone that looks like a plowshare. Joining with the ethmoid bone, it becomes the partition between the two nasal cavities, or septum. The lateral walls of the nasal cavity have two scroll-shaped bones called inferior nasal conchae. These thin, porous paired bones are elongated and curl upon themselves. Attached to the wall of the nasal cavity, the bones increase the amount of mucus membrane and olfactory nerve endings that contribute to the sense of smell.

Mandible, or Jawbone The largest and strongest bone of the face is the mandible. This horseshoe-shaped bone holds the lower teeth. Its parts include a curved, horizontal body and two perpendicular portions, the rami, that join the ends at right angles.

In adults, portions of the base are of equal depth and the rami are almost vertical, measuring from 110 to 120 degrees. However, in old age, the bone is greatly reduced in size. With the loss of teeth, some bone is absorbed and each ramus becomes oblique, with an angle that measures about 140 degrees.

Hyoid The U-shaped hyoid bone is unique in that it does not attach to any other bone. Located in the neck above the larynx, or voice box, it serves as the attachment for the muscles of the tongue.

The Vertebral Column, or Spine
The vertebral column is the backbone of the body, forming a supporting rod for the head, arms, and legs. The vertebral column is made up of a series of 26 bones called vertebrae (singular, vertebra). Cartilage and ligaments link the vertebrae together, allowing flexibility and giving support to the trunk. The column protects the spinal cord.

Vertebrae are grouped according to the region they occupy (Figure 11.6):

- Cervical, or neck—7 vertebrae

- Thoracic, or chest—12

- Lumbar—5

- Sacral—5 fused bones between the hip bones

- Coccygeal—tailbone region

In the embryo, 33 separate vertebrae exist. Before birth, the five sacral vertebrae and four coccygeal vertebrae fuse, forming a single bone at birth.

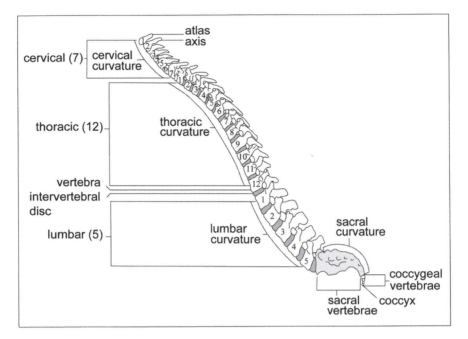

Figure 11.6   Spine. (Sandy Windelspecht/Ricochet Productions)

Seen from the side, the spinal column does not form a straight line. Several curves correspond to the sacral curves that appear before birth, and others develop later. The cervical curve in the neck appears when an infant begins to sit up and hold its head erect. Another forms in the lumbar region when the baby begins to walk. Changes in the curvature of the column result in shifts in the body's center of gravity. The average length of the spinal column in men is 28 inches (71 centimeters); in women, 24 inches (61 centimeters).

A Typical Vertebra   A typical vertebra has two parts (Figure 11.7):

- Body—the largest part, which is shaped like a short cylinder; this is the forward, or anterior, part

- Vertebral arch—a ring of bone formed by paired pedicles, or short, strong processes

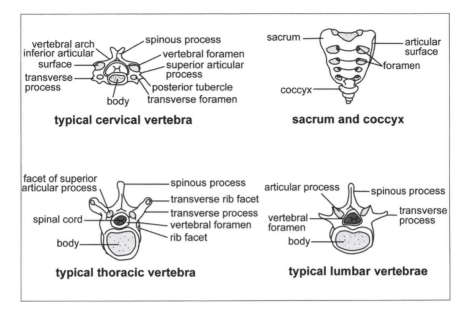

Figure 11.7 Typical vertebrae. (Sandy Windelspecht/Ricochet Productions)

Parts of the vertebral arch roof over an opening called the vertebral foramen, through which the spinal cord passes. Each vertebra has a spinous process for the attachment of ligaments and muscles of the back. Other processes permit the attachment of muscles and joints that connect individual vertebrae.

The following processes are on the vertebrae:

- *Pedicles*—short thick processes that project backwards, one on each side; vertebral notches are located above and below the pedicles

- *Laminae*—two broad plates directed backward and toward the middle

- *Spinous processes*—bony areas directed downward and backward from the junction of the laminae that serve as attachment sites for muscles and ligaments

- *Articular processes*—springing from the junction of the pedicles and laminae, their surfaces are coated with hyaline cartilage, or true cartilage that has a smooth and pearly appearance

- *Transverse processes*—located at each side of the joint where the lamina joins the pedicle; sometimes called wings, they also serve as attachments for muscles and ligaments

Because the vertebrae differ in subtle ways in their makeup, they are identified in the following ways:

- The letter C is for cervical; T for thoracic; L for lumbar.

- The vertebrae are numbered from top to bottom. For example, C2 is the second cervical vertebra; T3 is the third thoracic vertebra. Each number begins with the location in the new section.

Cervical Vertebrae   The seven cervical vertebrae in the neck are the smallest of the vertebrae. These vertebrae enable the head to turn in roughly three-quarters of a circle without moving the shoulder. Moving the eyes in conjunction with the neck generates almost a full circle of vision. Muscles run from the wings, or transverse processes, on the sides and to the rear of the vertebrae to the skull, shoulder blades, and lower vertebrae. These structures also steady the head on the neck.

Cervical vertebrae are different from thoracic or lumbar vertebrae in that they have a foramen, or hole, in each transverse process. There are also other differences:

- C1 is named for Atlas, the Greek mythological Titan who bore the weight of the world on his neck and shoulders. The atlas, or C1, allows the nodding movement. This bone has no vertebral body, only anterior and posterior arches. It joins with the odontoid process of C2, or axis.

- C2 is the second cervical vertebra, called axis. The axis allows movement from side to side. The odontoid process projects above the body and joins the anterior arch of the atlas. The dens is a process that forms a pivot around which the atlas rotates to allow head movement.

- C3–6 are vertebrae that are alike but without the special features of C1, 2, and 7.

- C7, called vertebra prominens, has the distinctive feature of a long spinal process that is thick and nearly horizontal.

Thoracic (Chest) Vertebrae   The thoracic vertebrae become larger from top to bottom to carry additional weight. These 12 vertebrae are intermediate in size between the cervical and lumbar vertebrae. They have the features of a typical vertebra, plus long slender spines that project downward. The 12 vertebrae join with the 12 pairs of ribs. The ribs join to shallow cups, or costal pits, on the body of the vertebrae. Ten thoracic vertebrae have two costal pits, or facets, on each side—one above and one below—to join the head of the same rib. This gives extra stability. The two sets move with every breath.

Lumbar (Lower Back) Vertebrae   These five vertebrae are the strongest and the largest because they support the weight of the body. The transverse processes and neural spine are thicker because they anchor the muscles that twist and level the lower back. Between the vertebrae is a cushion-like disk of cartilage. Although the lumbar vertebrae have features of a typical vertebra, they also have short, blunt spines that project to the rear. The lumbar spines do not overlap, so the area is a good place for a spinal tap—a drawing of spinal fluid for the diagnosis of certain conditions, such as meningitis, an inflammation of the covering of the brain called the meninges.

Acrum   The sacrum is a triangular bone formed by five fused vertebrae. The sacrum makes up part of the pelvis. The fused vertebrae sit like a wedge between other parts of the pelvis.

Coccyx   The end of the vertebral column is the coccyx, or tailbone. It results from the fusion of four coccygeal vertebrae, which are not as complex as the others in that they have no pedicles, laminae, or spine. The coccyx is made from four centers, one for each segment. The forming of the bone occurs in the following order during a person's lifetime:

- Between ages 1 and 4, the first segments ossify, or form bone.

- Between ages 5 and 10, the second unit forms bone.

- Between ages 10 and 15, the third unit forms bone.

- Between ages 14 and 20, the fourth unit ossifies.

- As people advance in age, the segments unite with one another.

- Later in life, especially in women, the coccyx often fuses with the sacrum.

Gymnasts demonstrate the amazing agility of the spine, which is most flexible during the younger years. As people age, knobs of bone grow on the vertebrae and cartilage disks between them become hardened.

Thorax, or Rib Cage

Because the lungs inhale and exhale, they need protection from being damaged. A solid case like the skull would not enable a person to breathe in and out; a better means of protection would be a group of moveable bars forming a flexible cage—just like the ribs. In fact, the ribs are closely spaced with ligaments and muscles in between. They move at the joints, with the spine and breastbone, making it possible for them to lift upward during inhaling and to move back down during exhaling.

The thorax, or rib cage, is made of bones and cartilage that protect the important organs of respiration and circulation. Its shape is like an inverted cone, with the narrow part above and the broad part below. From the back, the 12 thoracic vertebrae and the posterior part of the ribs form the cage. The sternum and costal cartilages form the front surface. The sternum is slightly convex and tilts downward and forward. The ribs form its side surface. Eleven intercostal spaces separate the ribs. The term costal means "pertaining to the ribs." The intercostal muscles and membranes are also found in this area.

A woman's thorax differs from a man's in the following ways:

- The woman's capacity is less.

- Her sternum is shorter.

- The upper level of her sternum is on a level with the lower part of her body.

- Her upper ribs are movable, allowing for a greater enlargement of the upper part of the thorax to accommodate expansion of the uterus during pregnancy.

Ribs   Twelve pairs of ribs are classified as typical, or true, ribs; false ribs; and floating ribs. All three types of ribs have the following structures:

- *Head*—the back and middle ends that join with demifacets, or cup-like structures, of two adjacent vertebral bodies

- *Neck*—a constricted region about 4 inches (2 centimeters) long that is beside the head

- *Tubercle*—next to the neck of the rib, this part joins to the transverse process of a vertebra

- *Body*—the shaft of the rib, the longest part of a typical rib

The ribs are described as follows:

- Ribs 1–7 are true ribs that attach directly to the sternum by means of costal cartilage and a true synovial joint (this joint has a clear lubricating fluid secreted by the synovial membrane, which will be detailed later in this chapter).

- Ribs 8–10 are false ribs joined via the costal cartilage of rib 7.

- Ribs 11–12 are floating ribs that do not articulate with the sternum or costal cartilage of the rib above (see photo).

Most men and women have 12 complete pairs of ribs. Sometimes a man or a woman will have 11 or 13 pairs, but this is unusual.

Sternum, or Breastbone
The sternum is a broad, flat bone forming the anterior wall of the thorax. Three parts make up its structure:

- *Manubrium*—the top part of the sternum means "handle," such as that of the handle of a sword.

- *Body*—the middle part articulates with the costal cartilages and ribs 2–7.

- *Xiphoid process*—the lower, or inferior, part of the bone varies in size and joins the bottom of the sternum; its name means "sword shaped."

Ligaments form the alignments to the clavicles, or collarbones, at each side of the top of the sternum. Costal cartilages are bars of true hyaline cartilage made of smooth, tough material that contribute to the elasticity of the rib cage.

The axial skeleton—composed of skull, vertebral column, ribs, breastbone, and pelvic and pectoral girdles—is the main part of the skeleton making up the head and trunk. These 80 bones protect the vital organs of heart, lungs, spinal cord, and brain. They form the foundation for the attachments of the appendicular skeleton.

Bones of the Limbs: The Appendicular System

Whereas the bones of the limbs had a mystical past, they serve today as the creative tools of modern existence Bones of the limbs and structures relating to them make up the appendicular skeleton. The word appendicular comes from the Latin, meaning "to hang from." Actually, the limbs do hang from the body as appendages.

The axial skeleton of skull, vertebral column, ribs, and sternum becomes the form to which the appendages are hung. Most bones do not lie in the body's central axis but in the extremities. The appendicular skeleton has 126 bones and includes the bones of the arms and legs and those of the shoulder and pelvic girdle.

Mammals, including humans, generally have legs longer than the torso. These bones account for great variation in height. For example, the length of the human spine varies little from woman to woman or from man to man. The woman's spine is basically 24 inches in length, and the man's is 28 inches. Differences in height result from the length of the leg bones. When a group of men are seated, they look the same height; but when they stand up, great differences in height are visible. Arm and leg, hand and foot—these appendages are similar, with the arrangement and number of bones being somewhat alike.

The importance of the human appendages is underlined by specialization. Quadripeds—animals that walk on all fours—have similar structures in all limbs. Apes and monkeys have hind feet that are very much like hands. In addition to opposable thumbs, these animals have opposable big toes that enable them to grasp. In humans, the limbs have more specialized jobs. The hands function to create and to work; the feet support.

The appendicular skeleton has 126 bones and includes the bones of the arms and legs, as well as those of the pectoral and pelvic girdles. The structures of the muscles, tendons, ligaments, and cartilage surround the bones at the joints and enable movement.

The Pectoral Girdle

The pectoral, or shoulder, girdle is composed of four bones: two scapulae and two clavicles. Usually a girdle is something that encircles as a complete ring. Looking down at the pectoral girdle from above, one can see a double crescent that encircles the upper part of the body. However, this girdle is incomplete, with the clavicles, or collarbones, being separated by the sternum in the front and a gap between two scapulae, or shoulder blades, in the back. In the back, the scapulae are connected only to the trunk by muscles. The bones allow for the attachment of muscles that firmly bind the arm to the trunk. These muscles also permit free movement of the arms.

Scapulae, or Shoulder Blades
Two pairs of bones in the upper torso connect to the bones of the arm. Located one in front and one in back, the shoulder blades literally float in a sea of muscles. The name scapula comes from a Greek word meaning "to dig." The shape is like a shovel or spade. Early humans used the scapulae of some animals as primitive digging tools. Shoulder blades are difficult to fracture and articulate with the clavicles.

The scapula—a flat, triangular bone that has two surfaces—forms the back part of the shoulder girdle. A spiny, prominent, shelf-like ridge extends obliquely across the blade. It supports the acromonium process. This is where the shoulder blade connects with the head of the humerus.

The scapula has two surfaces, three borders, and three angles. One angle, called the subscapular angle, appears to be bent in on itself along a line at right angles. This arched form strengthens the body of the bone while the summit of the arch supports the spine and the acromonium. Underneath the clavicle, the coracoid process of the scapula projects forward and serves as an attachment site for several muscles.

Clavicle, or Collarbone

The clavicles are located in front, ventral to the rib cage and just above the first rib. One can feel them at the base of the neck where the collar is located. The name clavicle is a Latin word that means "little key." To some people's imaginations, it looks like an old-fashioned key. The clavicle of birds is very important in folklore. It forms the familiar V-shaped wishbone: Break the long end, and your wish will come true.

The clavicle forms the anterior, or front, portion of the shoulder girdle. Some people describe this long, thin, curved bone as shaped like an S. Others say it is like the italicized lowercase f. Placed horizontally at the upper and anterior part of the thorax, it is immediately above the first rib. The upper surface is flat and rough and has impressions for attachments of the deltoid muscle that covers the shoulder prominence in front and the triangular-shaped trapezius muscle covering the back part of the neck and shoulders behind. In fact, the clavicle acts like a fulcrum that enables the muscles to give lateral motion to the arm. The part of the clavicle that joins the scapula is triangular. The other end joins a flattened projection of the scapula called the acromonium process. This process can be felt as the slight bony projection on the upper surface of the shoulder.

Women have shorter, thinner, less curved, and smoother clavicles than men. In people who perform manual labor, the clavicles are thicker, are more curved, and have prominently marked ridges for muscular attachments. The clavicle is the most commonly broken bone in the body because it transmits forces from the arm to the trunk.

The Arm

Hooked onto the pectoral girdle are the bones of the arm. Each arm consists of three principal bones: the humerus, ulna, and radius. Because of the joints in the bones, the arm is able to move. Also, arms are divided into three segments: upper arm, lower arm, and hand.

Humerus   The longest and largest bone of the upper arm is divided into a body and two extremities, or ends. This long bone of the upper arm extends from the shoulder to the elbow. The word humerus comes from the Greek word meaning "shoulder." Sometimes the humerus is called the brachium. However, it should not be confused with the word humorous, which means "comical" or "funny." The common term funny bone comes not from a bone, but from a nerve—the ulnar nerve—that passes over the elbow and creates a tingling sensation when hit.

The humerus has a smooth, rounded head that articulates with the scapula. On the lateral, or side, surface is a roughened area. Muscles and ligaments attach here. There are also several sites along the shaft of the bone where other muscles are attached. The bones of the forearm articulate on smooth surfaces at the lower end of the humerus.

Two knob-like projections—one on the lateral side and one in the middle—are attachment sites for the common extensor and the common flexor tendons that help move the forearm. Inflammation of the extensors causes the condition known as "tennis elbow." The distal end of the medial condyle of the humerus is called the trochlea, meaning "pulley." It articulates with the troclear notch of the ulna, which limits side movement and guarantees a hinge action.

Ulna   The ulna is the longer bone along the back of the forearm. The word comes from the Latin term that means "elbow." The end of the ulna is the bony portion of the elbow. At the upper end of the ulna is the olecranon process, a projection that fits into the olecranon fossa of the humerus when the arm is extended. These two bones form the joint at the elbow. On the lateral side of the ulna is a shallow notch for the head of the radius. The head of the ulna joins with the radius at the lower ends of the individual bones. The flattened surface at the lower end of the radius enables it to rotate around the ulna.

Radius   The word radius comes from the Latin word that means "ray," something that radiates outward from a center (originally, the word applied to the spokes of a wheel). Whoever named this bone thought this section of the lower arm seemed straight enough to be a spoke in a wheel.

The radius pivots on its long axis and crosses the ulna. Its proximal end has a smooth, rounded surface that articulates with the ulna. Next to the head is the neck; then comes the body, the long slender mid-portion also known as the shaft, or **diaphysis**. The bone ends with the styloid process, a projection that joins with the carpal bones of the wrist.

The radius and ulna form the forearm. When viewed from the correct anatomical position, the radius is on the side away from the body (lateral side), and the ulna is on the side toward the body (medial side). The ulna is longer than the radius and connects more firmly to the humerus. However, the radius is more involved in movement of the hand. A broken arm usually involves this bone.

The Hand

The hand has three parts: the wrist, palm, and fingers. Twenty-seven bones form this distal end of the limb. The large number of bones allows for the hand's versatility. A single bone would create a flipper-like, inefficient oar; a single line of bones would lead to bending as a unit, as in the spinal column. But the number of hand bones that spread out in two planes— length and width—introduce two-dimensional flexibility that permits delicate maneuvering. Thus the hand is a superb manipulative organ with four limber fingers and an **opposable thumb** that can act as pincer, grasper, twister, bender, puller, pusher—and manipulator of piano and computer keys.

Carpals The wrist, or carpus, has eight small bones. The Latin word carpus means "wrist." Ligaments hold these bones tightly together. To allow flexion, or movement, of the wrist, the carpal bones are arranged roughly in two rows. On their inner, or palmar, surface are attached some of the short muscles that move the thumb and little fingers.

Together, the bones of the wrist form a box-like structure. Each of the bones of the wrist has a special appearance and function. Early anatomists named them according to objects that were familiar at the time. The bones in the first, or proximal, row from side to middle are as follows:

*   *Scaphoid*—sometimes called navicular, from the Latin word that means "boat shaped." This bone is located on the floor of the anatomical wrist box. Hyperextension, or excess bending, of the wrist may fracture this bone.

- *Lunate*—from the Latin word meaning "moon shaped" or "crescent shaped"; the second carpal bone in the proximal row.

- *Triquetrum*—from the Latin word meaning "three cornered." The bone is the most medial in the row.

- *Pisiform*—from the Latin word meaning "pea shaped." This **sesamoid bone** is in a tendon and articulates with the triquetrum.

Bones in the second, or distal, row are as follows:

- *Trapezium*—from the Latin word that means "saddle" or "swing." This bone forms a saddle joint with the metacarpal bone of the thumb. Literally, the thumb swings on the trapezium.

- *Capitate*—from the Latin word meaning "head." This largest carpal bone is named for its rounded head. A punching blow with the fist generates forces that are transmitted through the third metacarpal bone to the capitate to the radius.

- *Hamate*—from the Latin word meaning "hooked." This describes the shape of the bone.

Metacarpals   The five metacarpal bones articulate with the bones of the wrist. The word metacarpal is from the Latin word that means "after the wrist." The other ends of the metacarpals are rounded and connect with the bones of the fingers. They are embedded in soft tissue and form the palms of the hand. The metacarpals are visible in the back of the hand. They are numbered 1 through 5 beginning with the thumb. Numbers 2 to 5 are almost parallel and do not move, whereas number 1 is set at an angle and has limited mobility.

The metacarpals are described as follows:

- Metacarpal 1—shorter and stouter than the others; diverges to a greater degree from the carpus; its surface is directed toward the palm

- Metacarpal 2—longest of the five bones; forms a prominent ridge

- Metacarpal 3—a little smaller than number 2

- Metacarpal 4—shorter and smaller than number 2

- Metacarpal 5—has only one facet on its base

The metacarpals articulate with the carpals and phalanges of the fingers.

Phalanges The 14 bones of the fingers are called phalanges. One of these bones is called a phalanx. In Greek history, the phalanx was a close formation of soldiers marching side by side and front to back. Fingers 2 to 5 have three phalanges decreasing in size from proximal to distal. The thumb has two phalanges.

Each finger has a formal name. The thumb is called pollex, from the Latin term meaning "strong." It is stronger than the others. (For example, people push a tack into a board with the thumb—hence the term thumbtack.) The second finger is the index finger, from the Latin word meaning "pointer." The middle finger is called medius, from the Latin for "middle." The fourth finger is the ring finger, or annularis, from the Latin for "ring." Last, as well as least, is the minimus, which means "least."

Fingers 2 to 5 have three phalanges, whereas the thumb has only two. The fingers move in and out or up and down. As each phalanx meets, it forms a hinge joint that enables part of the finger to bend. The thumb is more flexible than the other four fingers, with the distal end of its corresponding metacarpal bone being more rounded. Thus the thumb may cross the palm of the hand. This is why it is called an opposable thumb. Occasionally, small sesamoid bones are found within the tendons of the hand.

The Pelvic Girdle

Just as the pectoral girdle anchors the arms and hands, the heavier and stronger lower hip girdle supports the structures of the legs and feet. The word pelvis comes from the Latin word that means "basin." The pelvic girdle has two large hipbones attached to each other. In the back, these connect with the sacrum to make a bowl-shaped circle of bone. An oddity in the animal kingdom, humans' rounded pelvis is quite different from the elongated pelvis of other animals.

Hip Bone

It is a common mistake to think that the hipbone is the large ridge just below the waist. Actually, the hip is at the top of the thighbone and is hidden beneath heavy layers of muscle.

The hipbone is a large, flattened, irregular-shaped bone constricted in the center and expanded above and below. The prominent bones of the hip are called coxals or ossa coxae, from the Latin words meaning "hip bone." Sometimes the name is innominata, from the Latin word meaning "no name." The bones meet in the middle line in front and form the sides and anterior walls of the pelvic cavity.

Each hip bone is composed of three pairs of bones: the ilium, from the Latin for "groin"; the ischium, from the Greek word for "hip"; and the pubis, from the Greek word meaning "adult." The midline divides the bones, with one on each side. Together, they constitute the bony structure known as the hips.

Ilium   This fan-shaped bone forms the lateral, or side, prominence of the pelvis. Divided into two parts, the body of the ilium forms two-fifths of the acetabulum, the socket that fits the ball-shaped top of the thigh. The crest of the ilium can be felt on each side of the body just below the waistline. This is the bone most people mistake for the hip bone.

Ischium   This V-shaped bone forms the lower and back part of the hip bone. The body of this bone forms another two-fifths of the acetabulum. It is the site of attachments of many membranes and ligaments, including the hamstring, a major muscle in the groin. The muscles that enable one to sit are also attached to the ischium.

Pubis   This angulated bone forms the front part of the pelvis and makes up one-fifth of the acetabulum. It joins the ischium to form a pair of large holes. The holes, which are prominent in pictures of the skeleton, are called obturator foramina, Latin for "stopped-up holes." In reality, a membrane covers this area.

The term pubis is derived from the same root as the word puberty, describing sexual maturity. One of the signs of such sexual change is the growth of adult hair, or pubic hair. The bones located just under this region share this name.

In the front, the pubic bones are joined by cartilage similar to that in the vertebrae. This linkage is called the pubis symphysis, Greek for "growing together." In the back, the two bones that form the ilium do not meet but join the sacrum to form the sacroiliac joint, the articulation of the hip bones with the sacrum. Five sacral vertebrae form the sacrum, a connection so firm that sometimes the area is considered one bone—the sacroiliac. This area sometimes becomes a problem for human beings because of the stress of standing upright on two feet.

The sacrum and the two bones known as os coxae form the complete bony structure called the pelvis. The rounded, basin-like structure is found only in human beings. However, this arrangement is not perfect, as the pelvic basin tips forward and is not entirely upright.

During birth and early childhood, each coxal bone has the three separate parts of ilium, ischium, and pubis. But by age 20, these bones are firmly fused. The fusion takes place at the large cup-shaped cavity called the acetabulum, located near the middle of the outer surface of the bone. The acetabulum forms a socket, or depression, that gives it this name. The structure looked like a round cup the Romans used for vinegar, or acetum. Forensic scientists and anthropologists who examine skeletal remains use the hip for distinguishing between man and woman. In the woman, the bony ring of the pelvic girdle must be large enough for an infant to pass through during childbirth. The average lighter and thinner female girdle is two inches wider than that of the male. However, the rest of the woman's skeleton is much smaller and lighter. In the man, the bones are more massive, and the iliac crests are much closer.

Where the pubis bones meet at the **symphysis**, a disk of cartilage, in the woman, the bones form a right angle of 90 degrees, whereas the man's has an angle of 70 degrees. During childbirth, the fetus must pass through the pubic area. If the opening is too small, a problem may occur with birth.

By looking at skeletal bones, scientists can determine the number of children a woman has had because a record of the births is evident in the pelvis. Parturition (meaning "childbirth") scars begin to be deposited at about the fourth month of pregnancy when a hormone is released that softens the tendons that hold the pelvic bone together. These scars appear on the back side of the pubis symphysis.

The pelvic girdle, along with the sacrum, is a massive and rigid ring that is very different from the light and mobile shoulder girdle. But the pelvis, while sacrificing mobility, does provide strength and stability. The pelvic girdle supports the weight of the body from the vertebral column. It also protects the lower organs—including the bladder and the reproductive organs—and the developing fetus during pregnancy.

Thigh and Leg

Simply standing on a corner waiting for a bus is in fact an intricate balancing act. Some animals, such as bears, can stand on their back legs for a short period but will soon topple over. Humans, however, can stand upright for hours, leaving their hands free for other things.

The lower extremity is composed of the bones of the thigh, leg, foot, and patella, commonly known as the kneecap. Bones in the legs reflect a compromise with the upper part of the body: They are much stronger than the bones of the arm but are able to move less.

Femur   Between the hip and knee is a single bone called the thigh, or femur. It is the longest and strongest bone in the human skeleton, making up about two-sevenths of the person's height. Femur is from the Latin word meaning "thigh" (Figure 11.8).

The parts of this bone include the head, neck, trochanter, body, and patellar surface. The head of the femur, the proximal end, is rounded with ridges that anchor powerful leg muscles. It joins the ilium at an indented surface called the acetabulum, fitting into this rounded socket as a ball and socket. Below the head of the femur is the neck, a constricted area that is next to the head. Most of the blood supply to the head streams along this surface. The main shaft of the thighbone forms a wide angle. The neck region tends to become porous over time, especially in older persons, and therefore is a common site of fracture.

The lower end of the femur expands into a large flattened area with two bony processes on each side. Jutting out from the junction of the neck of the shaft is the trochanter. The greater trochanter is the insertion point

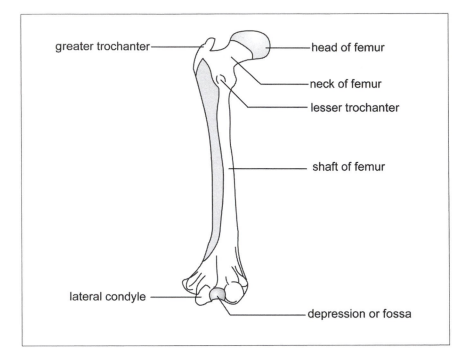

Figure 11.8   Femur. (Sandy Windelspecht/Ricochet Productions)

for several muscles, including the gluteus minimus. The lesser trochanter projects from the back middle surface and is an insertion point for major tendons and the muscles that form the posterior of an individual. The body is the long slender shaft that ends in the condyles, three rounded inferior ends that terminate at the kneecap.

Patella, or Kneecap   Differing from a similar structure in the arm—the elbow—the knee has a separate bone: the kneecap, or patella. The name of this small, flat, triangular bone comes from the Latin, meaning "small pan." The larger bone of the lower leg, the tibia, joins with both processes on each side. The larger bone of the lower leg joins with the tibia to form the knee joint. The kneecap protects this important joint, which is constantly pushed out ahead of the body as one walks or runs. Like the hyoid bone, the patella is a flat sesamoid bone located just in front of this joint.

It is not directly connected to any other bone. The patella develops in the tendon of the front thigh muscle. A ligament attaches the patella to the tibia.

When leg muscles are relaxed, the patella is not held in place by these muscles. The patella moves. Lodged in the tendon, it straightens the leg at the knee and increases the muscle's leverage while protecting the knee.

Tibia, or Shinbone Like the middle part of the arm, the leg has two bones. Whereas the bones in the arm are nearly equal in length, those in the leg are unequal. On the side nearest the body (medial side) is the larger of the two bones—the shinbone, or tibia. It is the second-longest bone in the body. The word tibia comes from the Latin word meaning "flute," which its length and shape resemble. The shinbone is long and skinny; it is slim where the stresses are least.

The tibia is the weight-bearing bone of the leg. It possesses heavy prominences, or condyles, that articulate with the femur to form the knee joint. In the front of the lower leg, the tibia can be felt as not very smoothly rounded and with a protruding ridge. At the distal end at the ankle, the heavy bony protuberance it forms at the inside of the ankle can be felt.

Fibula, or Calfbone This bone is much thinner than its partner, the tibia, and does not carry the body weight. It mostly anchors muscles that move the foot. The name comes from the Latin word meaning "pin." Its relationship to the tibia resembles the pin on the back of a brooch: well hidden. The fibula cannot be felt, as it is securely embedded in muscles. Sometimes it is called the splinter bone, as it resembles the splinter off the shinbone. The end of the fibula can be felt as the bony prominence on the outside of the ankle.

The Foot
The foot, or pes, consists of the ankle, instep, and five toes. Twenty-six bones make up the distal limb (Figure 11.9).

Tarsus or Ankle The ankle has a series of seven irregular bones, one less than the eight of the wrist. The word tarsus is from the Greek term meaning "wicker basket," apparently suggesting that the separate bones

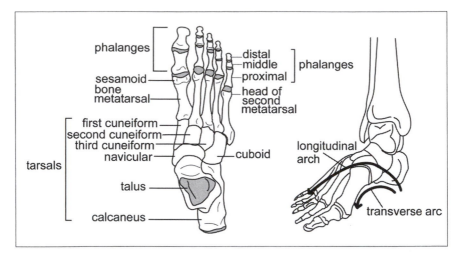

Figure 11.9   Foot. (Sandy Windelspecht/Ricochet Productions)

resembled the interwoven wicker strands of a basket. Bones of the tarsus are as follows:

- Calcaneus—from the Latin word meaning "heel." This largest tarsal, the strongest bone in the foot, extends back to form the heel. By extending backward, the two heels along with the two soles of the feet give humans **bipedal** support.

- Talus—from the Latin term meaning "ankle." This ankle bone articulates with the tibia, fibula, calcaneus, and navicular bones.

- Navicular—from the Latin word meaning "boat shaped." This bone articulates with the head of the talus and all three cuneiform bones.

- Cuboid—from the Latin word meaning "cube shaped." Soldiers in Rome used these bones, usually from horses, as dice for gambling.

- Cuneiform—from the Latin word meaning "wedge shaped." There are three of these bones.

As in the wrist, the bone structure enables the foot to move up and down in a single plane. The calcaneus helps support the weight of the body and serves as an attachment site for muscles of the calf of the leg. In fact, when standing, the body weight presses down through the talus and is divided evenly: Half travels to the heel bone, and the other half goes to the five remaining bones and the arch. Wearing high-heeled shoes upsets this balance, throwing the body weight to the balls of the foot.

Metatarsals The foot contains five metatarsals that join five sets of phalanges. The five metatarsals are similar to the metacarpal bones of the hand. Each of the metatarsals articulates with at least one of the tarsal bones, and sometimes with other metatarsals. The foot is shaped to form two main arches where the metatarsals join with the tarsal bones. One of the arches is longitudinal, lying perpendicular to the transverse arch. Together, they strengthen the foot and act as a spring to cushion certain movements. From any of a variety of conditions—including poor prenatal nutrition, excessive weight, fatigue, or incorrectly fitted shoes—lowered arches, or flat feet, can result. This problem causes unnatural stress and strain on the muscles of the foot and may lead to fatigue and pain while walking.

Phalanges The bones of the toes are similar in number to those of the fingers. Four of the toes have three phalanges, and the great, or first, toe has only two. Just like the fingers, the first digit of the big toe has two bones. The name for this bone is hallus, from the Latin word meaning "big toe." However, unlike the fingers, the toes have a sturdy, stout shape that bears weight. The bones help the foot push off, aiding in balance as they firmly grip the ground.

Joints, Ligaments, Tendons, and Cartilage

Not only is the skeleton a framework for the body, but it also constitutes a movable machine. Bones themselves are rigid, so the only possible motion occurs where two bones come together.

Indeed, the entire body is a complex interaction of matter and motion. A new field, biomechanics, merges the human machine and mechanics. Replacing joints diseased from arthritis, regrowing bone, and bone tissue engineering are hot research topics. These are only some of the current and future efforts to fathom the complex engineering of the human machine.

The sophisticated name for joint is articulation, from the Latin word meaning "to join." The study of the joints is known as arthrology, from the Greek word arthro, meaning "joint." The same root word is found in **arthritis**, meaning "inflammation of the joint."

Just because a joint is present does not imply mobility. Some joints allow no movement, some permit minor movement, and some allow free movement. The joints may or may not have ligaments that attach bone to bone or bone to cartilage.

Custom-designed according to the body part, joints are divided into three classes that describe the amount of movement between bones:

- **Synarthroses**, or immovable joints, permit no movement.

- **Amphiarthroses** permit only slight movement.

- **Diarthroses** are freely movable joints.

Tough **collagen** fibers called ligaments (from the Latin word that means "to tie") bind joints together and link bone to bone. Many are named from the two bones they attach. For example, the sphenomandibular ligament attaches the spine of the sphenoid bone and the mandible; the stylohyoid ligament connects the styloid process with the horn of the hyoid bone.

Ligaments and tendons are like rubber bands that hold the musculoskeletal system together. The material is very strong in resisting heavy loads. Although tendons and ligaments are made of dense, fibrous connective tissue, they differ in makeup and function. Ligaments bind one bone to another; tendons bind muscle to bone. Ligaments are 55 to 65 percent water. A special type of protein called collagen makes up 70 to 80 percent, with the protein elastin making up 10 to 15 percent of the dry weight (minus water). Tendons have 75 to 85 percent collagen and less than 5 percent elastin. Generally, tendons have fewer cells than ligaments do, and they are very sturdy. For example, the tendon of the foreleg of a horse can support the weight of two large automobiles. Sometimes a tendon is called by the Anglo-Saxon word sinew, which means "tough."

Immovable, or Fibrous, Joints

Immovable joints are called synarthroses, from two Greek words: syn, meaning "together," and arthro, meaning "joint." These joints are firm in their positions to prevent gliding or sliding. Examples are sutures in the skull.

Sutures are limited to the skull. Instead of cartilage between bones, fibrous tissue is located there. Necessary for skull growth, the joints are well marked in the young skull and barely visible in the aging skull. The only movement in the area is at birth, when cranial bones overlap to allow the baby to pass through the birth canal. Serrated little teeth that fit together join the suture. Later in life when growth is complete, they fuse. Suture joints in the skull are the following:

- Coronal sutures—the articulation between the frontal bone and the two parietal bones

- Intermaxillary suture—the midline of the hard palate that marks the line of the two palatine shelves

- Lambdoidal suture—the joint between the occipital and parietal bones that resemble the Greek letter lambda

- Metopic suture—a midline suture forming the articulation between two centers of the frontal bone

- Pterion suture—joins four bones: the greater wing of the sphenoid, the frontal, the parietal, and the squamous part of the temporal bone; this area is easily fractured with a blow to the side of the head

- Sagittal suture—joins the parietal and squamous portions of the temporal bone

Cartilage, or Amphiarthroses, Joints

These fibrous and cartilagenous joints occur where two bones are separated by a material that gives a little. Three such joints are synchondrosis, gomphosis, and syndesmosis.

Synchondrosis

This common type of joint is named from two Greek words meaning "join together with cartilage." The structure is a cartilage "sandwich," with bone as the "bread" on each side. The bone and cartilage fit together perfectly, and the whole joint is shaped like a cup. If movement occurs, the growing bone will be damaged, causing a condition known as a slipped epiphysis. (The **epiphysis** is the portion of a bone that is attached to another bone by a layer of cartilage.) A long pin can be inserted to hold the joint in place.

Gomphosis

These peg-and-socket joints occur between the teeth and jaws. The periodontal ligament holds this joint, which gives only a little. When the teeth bite down on a hard piece of candy, this joint absorbs the shock.

Syndesmosis

This type of joint is commonly known as a tight joint. In fact, tight ligaments limit many joints. An example is the inferior tibio-fibular joint between the two lower leg joints. A tight ligament limits the movement of this joint.

Symphysis is a word that describes two bones united by cartilage but designed to give a bit. For example, the symphysis pubis with ligaments and cartilage is normally closed, but female hormones signal it to open for childbirth.

The disks between the vertebrae are made up of fibrous cartilage and form a symphysis. These disks, which give just a little, are important shock absorbers between vertebrae. Joints between the disks of cartilage and other tissues permit some movement.

What Is Cartilage?

Cartilage comes in different types, each suited to a particular use:

1. Hyaline cartilage is the most prevalent type. It is found at the ventral ends of ribs, in the rings of the windpipe, and covering the joint surfaces of bones. The covering at the joint surfaces of bones is called articular cartilage.

2. Elastic cartilage is found in the external ear, the Eustachian tubes in the middle ear, and the epiglottis—the flap that covers the windpipe

during swallowing. Compared to hyaline cartilage, it is more opaque, flexible, and elastic. It is yellow and very dense.

3. Fibrocartilage occurs in the disks between vertebrae, in the pubis symphysis in the pelvis, and in the bony attachments of certain tendons. It may also form when hyaline cartilage is damaged.

Cartilage is mainly collagen embedded in a firm gel. Collagen is an albumin-like protein in connective tissue, cartilage, and bone. The material is more flexible than bone and lacks blood vessels. Cartilage cells receive nutrients from the diffusion of fluids from nearby capillaries (minute blood vessels that connect the smallest arteries and veins). Nose and ears are examples of cartilage that deteriorate at death. This is why a skeleton never has a nose or ears.

Synovial, or Diarthroses, Joints

Synovial joints are very different from fibrous joints in that they are constructed to allow a range of motion. Yet they are sturdy enough to hold the skeleton together. Cartilage covers bones near synovial joints so that ligaments may attach.

Constructed to give power and motion, the ends at the synovial joints have a thin but tough layer of articular cartilage. This clear material lessens friction and cushions joints from jolts. If the coating is destroyed, the bones grind against one another, producing a creaking sound. Between the bones and at the center of a synovial joint is the joint cavity that gives bones some freedom of movement. Synovial joints occur in a range of sizes and shapes (Figure 11.10).

Ball-and-Socket Joint

These joints are in the shoulder and hip. They allow for the freest movement. The surface of the epiphysis, or rounded head, of one bone fits in the cup-shaped socket of the other. For example, the ball of the femur fits tightly into the acetabulum of the hip bone. A rim of cartilage lines the socket and aids the firm grip on the femur. Some of the strongest ligaments in the body reinforce this joint. Because of the ball-and-socket joint, a person may move the leg in almost any position. **Dislocation**

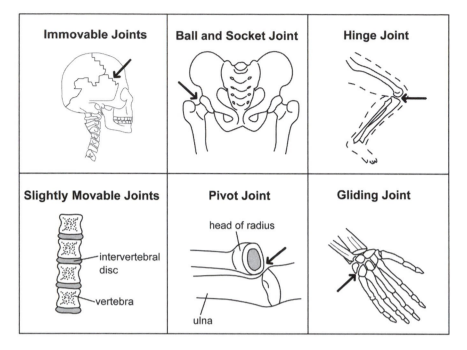

Figure 11.10 Joints. (Sandy Windelspecht/Ricochet Productions)

injuries are most common when the knee is flexed, as when sitting in a car and the impact from a collision causes the knee to strike the dashboard.

A similar ball-and-socket joint located between the humerus and scapula permits an even greater degree of freedom. The socket here is shallower than the one at the hip joint, enabling a person to turn the arm in a complete circle at the shoulder. For example, baseball pitchers doing a complex windup can move their arms in a full circle.

Hinge Joints
As a door swings back and forth on hinges in one plane only, so do bones connected at hinge joints. There are 40 hinge joints in the human body, including the elbow and knee.

The elbow allows motion in one plane only. This complex synovial joint consists of humerus and radius, humerus and ulna, and radius and ulna articulations—all within a common articular capsule. The proximal epiphysis of the ulna just fits between the two epiphyses of the distal end of the humerus, allowing movement back and forth but not from side to side.

More complex is the body's largest joint, the knee. This joint links the two rounded bulbs of the femur with the condyles of the tibia. The patella, or kneecap, covers this joint. In the act of walking, the knee is constantly rolling, gliding, and shifting orientation. To bear these enormous loads, a host of ligaments and tendons go into action. Two ligaments on each side of the joint prevent it from moving too far to one side. When torn, these ligaments usually repair themselves.

Not so with a second group of ligaments strapped across the joint, the cruciate ligaments. These **intracapsular ligaments** are the anterior and posterior cruciate ligaments. The anterior cruciate ligament, or ACL, is infamous in sports medicine. The role of these ligaments is to stop the joint from moving too far backward or forward. When the ACL is damaged, the knee is in great peril because it is extremely difficult to repair. Damage to the ACL has put many athletes out of their game.

The joint capsule of the knee has ligaments as well as menisci to add to stability. The medial and lateral menisci are crescent-shaped wedges of cartilage with a smooth, slippery surface. Menisci enable the joint to glide easily and to absorb the shock of daily activity. Menisci are the knees' weakest link, accounting for 90 percent of all knee surgeries. Muscles, such as the vastus medialis that hold the knee in place, also add to the stability of the region.

Other hinge joints occur between the first and second phalanges of the fingers and thumbs and between the second and third phalanges of the fingers. The same is true for joints in the toes. The lower jaw, or mandible, is mostly a hinge joint, but it can move from side to side in a rotary motion.

Pivot Joints

These joints provide rotary movement in which a bone rotates on a ring or a ring of bone rotates around a central area. In shaking the head "no," the movement between the first two vertebrae allows the turning of the skull on the spine.

Gliding Joints

These joints are found between the carpals of the hand and the tarsals of the feet. The joining parts slide over each other with angular or rotary motion. Similar joints are in the ribs and vertebrae.

Angular Joints

Where a football-shaped bone fits into a concave cavity, an angular joint occurs. This is found in the wrist, and it permits movement in two directions. Sometimes these are called condyloid or ellipsoid joints.

Saddle Joints

Similar to an angular joint in its range of movement, this joint is found only in the thumb. Each bone that forms a saddle joint has a concave and convex articular surface.

Plane Joints

These joints occur between two flat bones, where one moves horizontally over the other in both directions. They are found in the hand.

Synovial Fluid

When the bones move against each other, friction must be reduced. Several features work together to accomplish this. Portions of the bone are lined with a smooth layer of cartilage, and a capsule called the synovial capsule holds the bones together. This joint capsule permits movement and has great strength to prevent dislocation. Lining the joint is a membrane that secretes a lubricant called **synovial fluid**. Like egg white in texture, the fluid contains a substance called hyaluronic acid that helps to ease friction.

In these joints, there is also a membrane sac called a **bursa**. Bursae produce synovial fluid that bathes the end of the bone, allowing fluid movement. If the bursae become inflamed, a condition called bursitis causes severe pain.

Because bones are rigid, the only motion occurs where two bones come together as joints. Ligaments bind one bone to another; tendons bind muscle to bone. Although not all joints move, these structures at the joint allow the human being to move.

Summary

The human body's skeletal system is vital because it provides support to all of the other systems. Similar to the integumentary system, it also plays an important protective role. It protects the body's tissues and organs, as well as storing blood cells and other important nutrients. There are 206 bones in the skeletal system that are at once rigid to provide support and flexible to allow movement. One way of classifying bones is by shape. There are long bones, located in the body's lower extremities that are strong and light, and short bones that are located in the wrist and ankle. Flat bones are located in the ribs and the hip, while irregular bones make up the skull, face, vertebrae, and pelvis, and help to support weight. Sesamoid bones are short bones located within a tendon or joint capsule. Tendons, ligaments, and joints allow muscles and bones to move, working according to a lever system.

Another way that bones can be classified is through two systems: the axial skeleton and the appendicular skeleton. The axial skeleton includes the bones of the skull, vertebral column, ribs, and breastbone. The appendicular skeleton refers to the bones that hang from the axial skeleton, such as the collarbone, shoulder blades, arms, hands, thighs, and legs.

12

The Urinary System

Stephanie Watson

Interesting Facts

- About 25 percent of the total volume of blood in the body is pumped through the kidneys every minute.

- A normal, healthy adult kidney is about the size of a human fist and weighs about 5 ounces. But when cysts develop, as in polycystic kidney disease, the kidneys can grow to the size of a football and weigh upwards of 38 pounds.

- At birth, each kidney weighs about half an ounce. The kidneys do not reach their final weight until adolescence.

- The male urethra measures about 8 inches, compared with the female urethra, which may only be 1.5 inches long.

- A healthy adult bladder can comfortably hold 14–20 ounces (400–600 milliliters) of urine.

- The average adult ingests about 2.5 quarts of fluid per day and urinates between 1.5 and 2 quarts of urine each day.

- The average person urinates about 4–6 times per day. People who have severe cases of the chronic disorder interstitial cystitis may need to urinate up to 60 times a day.

- The average prostate weighs about an ounce and is about the same size and shape as a walnut. But in men who develop benign prostatic hyperplasia (BPH) as they age, the prostate can swell to the size of an orange.

- Each ureter stretches about 12 inches and, at its widest portion, measures about 0.5 inches around.

Chapter Highlights

- How the urinary system functions

- Components of the urinary system: kidneys, ureters, urinary bladder, urethra, prostate (men), and urinary sphincter muscles

- How the kidneys function

- How urine is formed

Words to Watch For

Acidosis

Adipose capsule

Aldosterone

Antidiuretic hormone

Bladder

Bowman's capsule

Calcitrol

Collecting duct

Cortex

Creatinine

Detrusor muscle

Distal convoluted tubule

Diuretic

External urethral sphincter

Glomerular capsule

Glomerular filtrate

Glomerulus

Hematuria

Hilius

Hypernatremia

Hyponatremia

Internal urethral sphincter

Ketone bodies

Kidneys

Loop of Henle

Major calyces

Micturition

Minor calyces

Mucosa

Papillary duct

Prostate

Proximal convoluted tubule

Renal capsule

Renal fascia

Renal pelvis

Renal pyramid

Renin

Semipermeable membrane

Serosa

Sodium/potassium	Urea	Uric acid
pump	Ureter	Urochrome
Sphincters	Ureteral orifices	
Trigone	Urethra	

Introduction

The human body is a sophisticated piece of machinery; its organs, nerves, muscles, and tissues are designed to perform one or more of the functions necessary to keep us alive and healthy. Within the body are several complex systems, each one specialized—each one crucial in its own way. The respiratory system supplies oxygen to the blood; the circulatory system transports that oxygenated blood throughout the body; the lymphatic system protects against infection; the endocrine system produces the hormones that direct bodily functions; the nervous system gathers, stores, and transmits sensory information; the skeletal and muscular systems maintain an upright posture and mobility; the reproductive system enables procreation; and the digestive system transforms food into energy.

Once the digestive system has completed its job and the body has taken the nutrients it needs, waste products are left behind in the blood. If allowed to build up, these poisonous wastes would eventually destroy cells, tissues, and organs, causing the body to simply shut down. Fortunately, humans have a built-in mechanism, called the urinary system, to rid the body of wastes. The urinary system turns toxic materials into urine, stores and carries that urine, and removes it safely from the body.

Waste removal is the urinary system's primary responsibility, but it has other important functions as well. For example, the system must maintain the proper balance of water and chemicals—ensuring that the body is hydrated but not drowning in fluid. It does this, in part, by controlling the amounts of **electrolytes**—inorganic compounds such as sodium, potassium, magnesium, and calcium—that conduct electric currents and regulate the flow of water molecules across cell membranes. Finally, the urinary system continuously monitors and regulates the acidity of body fluids.

The first section of this chapter delves into the inner workings of the urinary system. Included are detailed descriptions of each organ—kidneys, ureters, bladder, urethra—and their collective role in urine production, storage, and elimination. This includes the urine pathway from kidney filtration to bladder storage and finally to removal via the urethra. This section also describes in detail the process by which the kidneys filter waste products out of the blood and return necessary electrolytes and nutrients to the body, how the filtered urine travels to the bladder for storage, and how a combination of voluntary and involuntary nerves work together to release the urine from the body.

Parts of the Urinary System

To perform all of these sophisticated functions requires a collaboration of organs, tubes, muscles, and nerves. The primary components of the urinary system are the kidneys, ureters, urinary bladder, urethra, and urinary sphincter muscles.

The Kidneys

The two bean-shaped **kidneys** are the functional core of the urinary system (see Figure 12.1). They keep the body free from impurities, maintain a healthy water and chemical balance, oversee the composition of electrolytes, regulate blood pressure, and secrete several important hormones.

The kidneys are located on either side of the spine toward the back, just underneath the rib cage (Figure 12.2). The right kidney is slightly lower than the left to make room for the liver. In an average adult, each kidney measures about 5 inches long, 3 inches wide, and 1 inch thick, and weighs about 5 ounces. Three layers of tissue encase and protect each kidney: The **renal** (renal is another word for kidney) **capsule**, a smooth fibrous membrane, forms the innermost layer. It is surrounded by the **adipose capsule**, a layer of fatty tissue. Finally, the outermost layer, the **renal fascia**, is composed of connective tissue that holds the kidney to the abdominal wall.

The outer portion of the kidney is called the **cortex**. In the center of the kidney is the **medulla**, which contains 10–15 cone-shaped collecting ducts called **renal pyramids**. The renal pyramids drain urine into cup-shaped receptacles called **minor calyces**. From here, the urine flows into

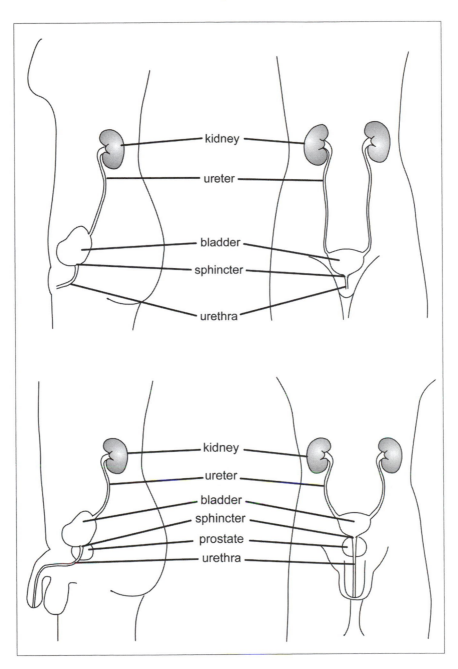

Figure 12.1 The Female and Male Urinary Systems. The two ureters lead from the kidneys to the bladder. The female urethra is shorter (1 ½ inches) than the male urethra (8 inches). In the male, the walnut-shaped prostate encircles the urethra. (Sandy Windelspecht/ Ricochet Productions)

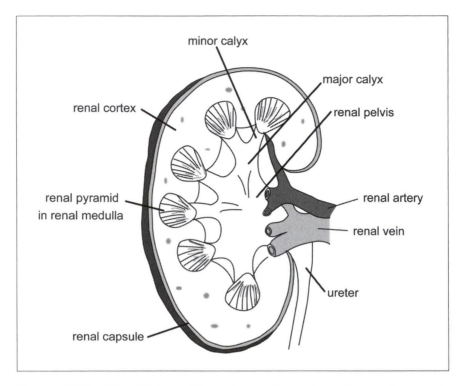

Figure 12.2   The Kidney. Filtered urine drains from the renal pyramids into the minor calyces, through the major calyces, and on into the renal pelvis. From there, it flows down through the ureter to the bladder. (Sandy Windelspecht/Ricochet Productions)

larger openings called **major calyces**, through the funnel-shaped renal pelvis, and on to the ureter and bladder.

Most kidney function actually takes place in microscopic cup-shaped capsules called nephrons. Each kidney contains about one million **nephrons**, and it is here that the blood is filtered. Toxic wastes are removed while water and necessary nutrients are reabsorbed into the system. Nephrons also control the blood **pH** level—in other words, they make sure that the blood is neither too acidic nor too alkaline. Inside each nephron is the **glomerulus**, a network of tiny blood vessels or capillaries, which are contained in a thin sac called the **glomerular capsule** (also called the **Bowman's capsule**). About 25 percent of the total blood pumped by the heart passes through the kidneys every minute. The blood enters each kidney via the renal artery in the **hilius** (the curved notch near the center on the side of each kidney), then branches off into the capillaries of the glomerulus. As the blood flows out of the glomerulus, it passes through a three-layered membrane, beginning the filtration process. The membrane stops blood cells and large protein molecules from passing through, but allows water, electrolytes, sugars, and amino acids to continue into the glomerular capsule.

When the filtered fluid, called **glomerular filtrate**, reaches the glomerular capsule, it moves through a small coiled tube—the **proximal convoluted tubule** (PCT). It is in this tubule that the real recycling work begins. Water, sodium, sugar, calcium, proteins, and other substances the body needs are reabsorbed into the bloodstream through tiny capillaries. The remainder of the fluid, containing substances not reabsorbed by the blood—such as water, **urea** (a waste product created by the breakdown of proteins), and excess salts—is carried through the U-shaped **loop of Henle**, to the **distal convoluted tubule** (DCT). Several distal tubules empty the waste, called urine, into a single **collecting duct**. The collecting duct in turn empties urine into larger **papillary ducts**. As it moves down the path from proximal convoluted tubule to papillary duct, the solution is continually being filtered, so that by the time it reaches the ureters, about 99 percent of the original glomerular filtrate has been reabsorbed into the bloodstream. The cleansed blood makes its way back to the heart through the renal vein, while the urine flows into the calyces of the renal pelvis on its way to the ureters.

Kidney Functions

The kidneys have three main functions: homeostasis, waste removal, and hormone secretion.

Homeostasis: The primary function of the kidneys is **homeostasis**—maintaining a balance of fluids within the body. The body contains more than 40 quarts (37 liters) of fluid, which is found in and around the cells. About two-thirds is intracellular fluid, located within the cells themselves. About 75 percent of the remaining extracellular fluid is found in the tissue outside of the cells (called interstitial fluid), and the other 25 percent is contained in plasma, the fluid portion of blood.

Water passes in and out of these three fluid areas via a process known as **osmosis**. Surrounding each cell is a **semipermeable** (or selectively permeable) **membrane**, which separates fluids of different concentrations. The semipermeable membrane allows certain molecules to pass through while restricting the movement of other molecules. In osmosis, water moves across this membrane via a passive process called **diffusion**, from an area of higher concentration to an area of lower concentration until the two volumes are equal. The process is called passive because fluid is not pushed across the membrane by any outside force, but simply flows from higher to lower concentration.

The body takes in about 2.5 quarts (2,500 millimeters) of water every day through food and beverages. What goes in must equal what goes out, and the body has several routes by which fluid can exit the body: the kidneys (urine), skin (perspiration), lungs (breath), and intestines (feces). When a malfunction in the water removal process occurs, the body becomes overly saturated or parched. Too much water in the blood can force the heart to work harder and dilute essential chemicals in the system. Dehydration, or too little water, can lead to low blood pressure or shock and is potentially fatal. The kidneys help to balance the fluid in the body by reabsorbing liquid into the bloodstream when levels get too low, or by eliminating excess fluids when levels rise too high. These processes are overseen by the hypothalamus, the part of the brain that also regulates metabolism, body temperature, blood pressure, and hormone secretion.

If the concentration of water drops too low (because not enough liquid was ingested or because fluid was lost through sweating, vomiting, or diarrhea), neurons called **osmoreceptors** send a message to the hypothalamus, which in turn tells the pituitary gland to secrete antidiuretic hormone (ADH; also known as vasopressin) into the bloodstream. This hormone increases the permeability of the distal convoluted tubules and the collecting ducts in the nephrons of the kidneys, thus returning more fluid to the bloodstream. When more water is reabsorbed, the urine becomes more highly concentrated and is excreted in smaller volume. When the fluid concentration in the body is too high, ADH is not released. The distal convoluted tubules and collecting ducts are less permeable to water, and the kidneys filter out excess fluid, producing a larger volume of more dilute urine.

The kidneys must also maintain a balance of sodium, potassium, and other electrolytes in body fluids. To do this, they separate ions from the blood during filtration, returning what is needed to the bloodstream and sending any excess to the urine for excretion. Electrolyte levels are directed by the endocrine system, a collection of hormone-releasing glands. Hormones are chemical signals that travel through the bloodstream, triggering cells to complete a particular job.

Sodium and potassium are two of the most important electrolytes, because without them, fluids would not be able to properly move between the intracellular and extracellular spaces. Sodium is the most abundant electrolyte in the extracellular fluid, and it also plays an important role in nerve and muscle function. The presence of too much sodium (a condition called **hypernatremia**) will cause water from inside the cells to cross over into the extracellular region to restore balance, causing the cells to shrink. If nerve cells are affected, the result can be seizures, and in rare cases, coma. Too little sodium (called **hyponatremia**)—lost from excessive diarrhea, vomiting, or sweating—can send water into the cells, causing them to swell. This can lead to weakness, abdominal cramps, nausea, vomiting, or diarrhea. The swelling is even more dangerous if it occurs in the brain, where it can lead to disorientation, convulsions, or coma.

Potassium assists in protein synthesis and is crucial for nerve and muscle function. Too little potassium can lead to a buildup of toxic substances in the cells that would normally pass into the extracellular fluid. To prevent a sodium-potassium imbalance, the cells use a mechanism called the **sodium/potassium pump**. This pump is a form of active transport (as opposed to the passive transport used in osmosis), which means that fluid can pass from one side of a semipermeable membrane to another, even if the concentration is already high on that side. But active transport requires energy to push molecules across the membrane. That energy is derived from adenosine triphosphate (ATP), a byproduct of cellular respiration. Once activated by ATP, the sodium/potassium pump pushes potassium ions into the cell while pumping sodium ions out of the cell until a balance is reached.

Endocrine hormones regulate the amount of sodium and potassium in the bloodstream. In the case of a sodium imbalance, an enzyme secreted by the kidneys, called **renin**, stimulates the production of the hormone aldosterone by the adrenal glands located just above the kidneys. Aldosterone forces the distal convoluted tubules and collecting ducts in the nephrons to reabsorb more sodium into the blood. It also maintains potassium homeostasis by stimulating the secretion of potassium by the distal convoluted tubule and collecting ducts when levels in the bloodstream get too high. Aldosterone also indirectly regulates the balance of chloride. As sodium is reabsorbed, chloride is present and is passively reabsorbed into the bloodstream.

Parathyroid hormone (PTH), produced by the four parathyroid glands in the neck, regulates levels of bone-building calcium and phosphate. When calcium concentrations in the body drop, PTH pulls calcium from the bones, triggers the renal tubules to release more calcium into the bloodstream, and increases the absorption of dietary calcium from the small intestine. When too much calcium circulates in the blood, the thyroid gland stimulates the production of another hormone, **calcitonin**, which causes bone cells to pull more calcium from the blood and increases calcium excretion by the kidneys. PTH decreases phosphate levels in the blood by inhibiting reabsorption in the kidney tubules, and calcitonin stimulates the bones to absorb more phosphate.

In addition to fluid and electrolyte balance, the kidneys play a crucial role in regulating the acidity, or pH, of fluids in the body. Water in the body is composed of hydrogen and oxygen molecules, which are held together by a chemical bond. Often the hydrogen and oxygen molecules separate into the positively charged H^+ ions and the negatively charged OH^- ions. An excess of H^+ will make the solution acidic, and too much OH^- produces an alkaline solution. Acidity is generally measured on a scale of 0 to 14. A neutral solution measures 7, right at the center of the scale. The higher the pH, the more alkaline the solution; the lower the pH, the more acidic the solution. The body pH must remain within a very narrow range, between 7.35 and 7.45, in order to survive. Fortunately, when the body fluids become too acidic or too alkaline, the kidneys either eliminate or reabsorb hydrogen ions until the pH returns to within its normal range.

Waste Removal: As food moves through the stomach and intestines, digestive enzymes break the nutrients into smaller particles to be used by the body. This breakdown process releases several toxic waste products into the bloodstream. These include:

- *Urea*: Amino acids, derived from protein metabolism, are broken down in the liver to form ammonia. Because ammonia is too poisonous for the body to process, the liver converts it into the less toxic urea for removal.

- *Uric acid*: This is formed by the breakdown of purines (components of foods) in the tissues.

- *Ketone bodies*: These are produced by the breakdown of excess fatty acids in the liver.

- *Creatinine*: This is a by-product of muscle metabolism.

If any of these wastes were allowed to build up in the blood, they would eventually poison the blood and cells. The kidneys filter out dissolved wastes from the bloodstream to form urine, which is eventually removed from the body. More on urine formation will be detailed later in this chapter.

Hormone Secretion: The kidneys either secrete or activate three essential hormones:

- *Erythropoietin*: This stimulates the production of red blood cells in bone marrow.

- *Calcitrol*: This promotes bone growth by increasing the levels of calcium and phosphorous in the blood.

- *Aldosterone*: This regulates blood pressure and sodium balance by increasing the filtration of blood in the kidneys, increasing water reabsorption, and decreasing the amount of sodium that is lost. The kidneys do not actually produce aldosterone, but they do control its production by secreting renin, an enzyme that converts a protein in the blood called angiotensin to angiotensin I. As it passes through the lungs, angiotensin I is converted into angiotensin II. Angiotensin II stimulates the release of the hormone aldosterone from the adrenal cortex.

The Ureters

Out of each kidney extends a **ureter**, a thin, hollow tube that reaches down into the bladder. Each ureter stretches about 12 inches and, at its widest portion, measures about 0.5 inches around. The ureters pierce the bladder walls from either side, forming a U shape. At the bottom of the U, the ureters connect to the triangular-shaped area on the bladder floor called the trigone.

As the kidneys turn waste into urine, muscles lining the ureter walls help to push the urine down into the bladder for storage. Urine enters the bladder through openings called **ureteral orifices**. Valve-like mucous membranes inside the ureters keep the urine inside the bladder and prevent it from traveling back up toward the kidneys where it could cause an infection.

The Bladder

The hollow, muscular **bladder** stores urine until it is time for elimination (Figure 12.3). The bladder is located in the abdomen, just behind the pubic bone. As mentioned in the previous section, the ureters pierce the top of the bladder at a diagonal angle, forming a U shape where they intersect with the trigone. At the bottom of the trigone, in the neck of the bladder, is the opening to the urethra through which urine exits the body.

The inside of the bladder is composed of three layers: the **mucosa**, **detrusor muscle**, and **serosa**. The serosa, or outer coat, is made up of fibrous tissue. The detrusor muscle is actually a collective term for three layers of smooth muscle. This muscle is involuntary, meaning that it is not consciously controlled but is under the direction of the autonomic nervous system (see the section on urine removal later in this chapter for a description of the autonomic nervous system's role in bladder contractions during urination; for more information on the autonomic nervous system, see Chapter 8 in this encyclopedia). Finally, the mucosa, or inner lining, protects the bladder from infection.

As the bladder fills with urine, it stretches like a balloon. A healthy adult bladder can comfortably hold 14–20 ounces (400–600 milliliters) of urine. When the bladder fills up, a message is sent to the spinal cord indicating the need to urinate. The bladder muscles are relaxed during filling, but they contract during urination to push the urine down the urethra and out of the body. As the bladder contracts, its walls compress, preventing urine from

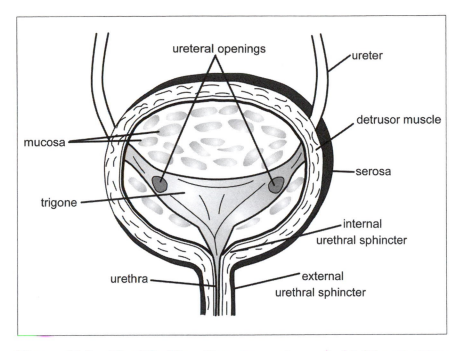

Figure 12.3   The Bladder. The ureters pierce the bladder on either side, reaching down into the organ to form a triangular shape, called the trigone. As urine funnels down the ureters and into the bladder, the detrusor muscle in the bladder wall is relaxed. When the bladder empties, the detrusor muscle contracts, and the internal and external urethral sphincters relax. (Sandy Windelspecht/Ricochet Productions)

backing up into the ureters and kidneys. The angle of the bladder also shifts during urination. While it is filling, the bladder is tilted at the point where it attaches to the urethra to prevent leakage. But during urination, the angle shifts to allow urine to flow down the urethra and out of the body.

The Prostate

Only men have a **prostate**, the doughnut-shaped gland that surrounds the urethra at the point where it connects to the bladder (Figure 12.4). In young men, the prostate is about the size and shape of a walnut, but with increasing age it can enlarge to about the size of an orange.

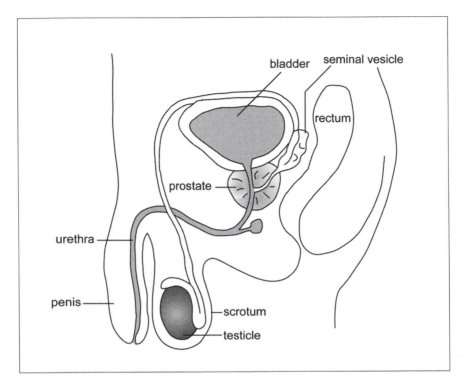

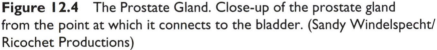

Figure 12.4 The Prostate Gland. Close-up of the prostate gland from the point at which it connects to the bladder. (Sandy Windelspecht/ Ricochet Productions)

The primary function of the prostate is to add nutrients and liquid volume to the sperm. Thousands of tiny glands inside the prostate produce a fluid that is mixed with sperm in the urethra during orgasm. The combined fluids, called semen, are released through the penis during ejaculation (the reproductive functions of the prostate are discussed in greater depth in Chapter 8 of this encyclopedia). The prostate also protects the bladder from infection. Its muscular fibers squeeze to help control the flow of urine into the urethra.

The Urethra

The thin, muscular tube of the **urethra** connects the bladder to the exterior and provides a passageway for urine to leave the body. The wall of the urethra is made up of a mucous membrane, as well as a layer of smooth muscle tissue. The length, path, and function of the urethra differ in men and women.

In men, the urethra is about 8 inches long and is connected from the bladder to the tip of the penis. It not only carries urine, but it also serves as the duct through which semen is released during ejaculation. In women, the urethra only measures about 1.5 inches in length, which is why women are much more likely to suffer a urinary tract infection than men. This short length makes it easier for urine to slip back up into the bladder and cause infection. The urethra connects to the top of a woman's vagina, just beneath the clitoris, and its sole function is to carry urine out of the body.

The Urinary Sphincter

Two groups of muscles, called **sphincters**, control the flow of urine out of the bladder. Where the bladder and urethra meet is a ring of smooth muscle called the **internal urethral sphincter**. These involuntary muscles stop urine from flowing back up the urethra to the bladder. At the end of the urethra is the **external urethral sphincter**, voluntary muscles that release and tighten to start and stop urine flow. When open, the internal and external sphincter muscles allow urine to exit the body. When closed, they prevent the urine from escaping.

The sphincter muscles are designed to work in conjunction with the bladder. When the bladder relaxes to allow urine to enter, the sphincter muscles remain closed to prevent leakage. As the bladder contracts during urination, the sphincter muscles relax, allowing urine to flow out of the body. If they fail to work together, urine may leak from the bladder, leading to incontinence.

What Is Urine and How Is It Formed?

Every day, about 190 quarts (180 liters) of blood plasma pass through the nephrons of the kidneys. Most of that liquid—about 188 quarts—is cleansed and returned to the bloodstream. The remainder is removed from the body as urine.

Urine formation involves three processes:

1. *Filtration*: Blood from the renal artery enters the nephrons of each kidney and passes through the tiny filtering units of the glomeruli. Blood pressure inside the capillaries of each glomerulus pushes water, small molecules (such as glucose, amino acids, and waste products like urea), and electrolytes into the Bowman's capsule, while leaving the larger blood cells and proteins in the bloodstream. The filtered fluids are called the glomerular filtrate. The amount of filtrate that forms in the glomerular capsule of both kidneys every minute is called the glomerular filtration rate (GFR). In a healthy adult, the normal GFR is about 4.2 ounces (125 milliliters) per minute.

2. *Reabsorption*: When the glomerular filtrate exits the glomerular capsule, it enters the proximal convoluted tubules. Nearly all of the water, sodium, and nutrients such as glucose, potassium, and protein are reabsorbed from the filtrate into the bloodstream. About half of the urea is also reabsorbed; the rest is excreted in the urine. The movement of these substances occurs through active and passive transport. Glucose, amino acids, sodium, and potassium are carried across the cells of the tubules via active transport. Water is reabsorbed passively via osmosis.

3. *Secretion*: As the filtrate enters the loop of Henle and moves up the distal convoluted tubule, wastes from the blood are added. Secretion is essentially reabsorption in reverse—rather than substances moving from the filtrate into the blood, they move from the blood into the filtrate. Substances added to the filtrate may include excess hydrogen, potassium, nitrogenous wastes (urea, creatinine, and uric acid), and certain drugs (such as penicillin). After rising up the ascending limb of the loop of Henle, the filtrate moves through the distal convoluted tubule and into the collecting duct, where it is referred to as urine.

What Makes Up Urine?

Because the human body is composed of about 50 to 70 percent water, it makes sense that the urine is made up primarily of water. In fact, about 95 percent of urine is water. The remainder is made up of dissolved wastes, including urea, creatinine, uric acid, and ketone bodies. The urine may also contain small amounts of substances the body normally uses, like sodium, potassium, and calcium. If the bloodstream contains excessive amounts of these nutrients, the kidneys will excrete the leftover portion into the urine. Other substances may end up in the urine that signal a problem within the body. The presence of protein or white blood cells during a urinalysis may indicate an infection or inflammation of the kidneys. Glucose in the urine may signal diabetes.

The volume and concentration of water in the urine are determined by how much water is reabsorbed as the filtrate passes through the end of the distal convoluted tubules and collecting ducts on the final leg of the urine production process, as controlled by the **antidiuretic hormone** (ADH) (see the section "How Is Urine Removed from the Body?" later in this chapter). If the body is dehydrated, more ADH is produced, making the cells more permeable to water. As more water is reabsorbed by the kidneys, the urine becomes more concentrated. If there is too much water in the body, less ADH will be produced, and water will not be reabsorbed in the distal convoluted tubules and collecting ducts, making the urine more dilute.

What Are the Characteristics of Urine?

Urine is typically yellow or amber in color. The yellow color is created by **urochrome**, a pigment produced from the breakdown of bile (the yellowish fluid secreted by the liver to digest fats). The color of urine may change depending on fluid levels in the body and on the types of foods ingested. A deficiency of fluid in the body forces the kidneys to absorb more water, creating a more concentrated, darker-colored urine. Too much fluid in the body results in a more dilute, lighter-colored urine. If the urine turns green or red, it may simply be because a person's diet included asparagus, beets, foods containing dyes, or certain medications. A change in color alone is usually no cause for concern, but the presence of pinkish or red blood in the urine (**hematuria**) can be a sign of disease (like kidney stones) or infection.

Normal, healthy urine is clear immediately after urination, but becomes cloudy when left standing. Likewise, urine does not have a strong smell as it leaves the body, but when left outside for any length of time, it develops an ammonia-like odor. This occurs because of the process in which the liver breaks down proteins. Referring back to the discussion on waste removal in the kidney function section of this chapter, the breakdown of proteins releases ammonia, which is too toxic for the body to handle. Ammonia is converted into the less toxic urea before it is sent to the kidneys for filtration. But when urea in the urine is exposed to oxygen in the air, it converts back to ammonia, which explains the strong odor. If the urine has a strong, foul smell immediately after leaving the body, it may be the result of bacteria from a urinary tract infection. In diabetics, urine will have a sweet, fruity smell because it contains excess ketone bodies.

The normal pH of urine ranges from 4.5 to 8.0, averaging around 6.0. The more acid the body retains, the more acidic the urine. Certain conditions may lead to overly acidic urine, including uncontrolled diabetes, diarrhea, dehydration, and **acidosis** (an abnormal increase in acidity when blood pH drops below 7.35). Urine tends to be more alkaline when an individual suffers from urinary tract obstruction or chronic kidney failure. A vegetarian diet can also make the urine more alkaline.

How Is Urine Removed from the Body?

The average person drinks between 1.5 and 2 quarts (1,500–2,000 milliliters) of fluid per day, and voids between 1 and 2.5 quarts (1,000–2,500 milliliters). Obviously, the more liquids that are ingested, the more urine the kidneys produce and eliminate. How much urine is produced also depends on the types of liquids ingested. For example, soda and coffee contain caffeine, which is a **diuretic**. A diuretic increases urine production and leads to more frequent urination than a nondiuretic like water.

If an individual urinates more than the volume of liquids ingested, the body becomes dehydrated. Prolonged dehydration can be caused by excessive sweating, vomiting, diarrhea, or extremely rare conditions like diabetes insipidus. If the fluid level in the body drops too low, serious illness may occur. Extreme dehydration can eventually result in death. Thankfully, the body has a built-in regulating system to prevent dehydration. When the level of fluid in the extracellular spaces drops, receptors in the hypothalamus of the brain release the antidiuretic hormone (ADH) into the bloodstream. ADH increases the reabsorbtion of water in the distal convoluted tubules and collecting ducts of the kidneys, creating a more concentrated urine. It also triggers that dry-mouthed feeling of thirst.

The average person urinates every 3–5 hours throughout the day and can sleep through the night without having to use the bathroom. Of course, what is normal varies from person to person. Some people may drink more, and therefore find it perfectly normal to urinate every 1–2 hours, while others drink less or have more accommodating bladders, and may only have to go every 5–6 hours. The volume of each urination can vary from about 4 ounces to 34 ounces, depending upon how much the kidneys are producing and how much the bladder can comfortably hold.

The process by which urine is released from the bladder, through the urethra, and out of the body is called **micturition**, or urination, and it occurs continuously. Every 10–15 seconds, small amounts of urine are forced down the thin tubes of the ureters and into the bladder. The bladder stores urine until an appropriate time—that is, until the person has reached a bathroom. As mentioned earlier, when the bladder fills, its muscles relax, but the muscles of the sphincters that surround the bladder opening tighten to prevent leakage. The process leading up to urination is controlled by the central nervous system and involves a combination of involuntary and voluntary nerve impulses. The bladder is controlled by the autonomic nervous system, which oversees other involuntary actions such as breathing. The voluntary muscles of the urethra are controlled by the somatic nervous system. As the amount of urine rises and the bladder stretches, receptors in its walls send a message to the spinal cord via sensory neurons that it is getting full. The spinal cord passes the message along to the brain in the form of impulses, and the result is that feeling of pressure in the lower abdomen that signals the need to urinate.

When it is time to urinate, parasympathetic nerves release the neurotransmitter acetylcholine, which causes the detrusor muscle in the wall of the bladder to contract and the internal urethral sphincter to relax. At the same time, the brain tells the sphincter muscles, which until now have been tightly holding the bladder and urethra shut, to relax. As the muscles relax, the urethra opens, and urine is allowed to exit the body.

When the Urination Process Fails

Losing voluntary control over the micturition, or urination, process is called incontinence. Babies naturally wet themselves because their brains and spinal cords have not matured enough to control the sphincter muscle. But if incontinence occurs in adults, it is usually the result of disease or injury to the nerves controlling the bladder, damage to the external sphincter, or infection.

The opposite problem, retention, is the inability to urinate. This may be caused by a blockage in the urethra or bladder neck, an uncontrolled contraction in the urethra, or the lack of an urge to urinate.

Summary

Each of the human body's systems performs a significant job that keeps it functioning and healthy. As emphasized throughout this encyclopedia, each system depends on the other systems to perform its duties. The urinary system's primary responsibility is to remove waste from the body. However, this system has other important jobs also, including ensuring that the body is properly hydrated and the body fluids are in proper balance. The organs or components of the urinary system that are integral to its function are the kidneys, ureters, bladder, urethra, and in men, the prostate. These organs play the key roles in urine production, storage, and elimination as well as the other processes of the urinary system.

Glossary

Abduction Withdrawal of a part of the body from the body's axis.

ABO group The name of the genetic system that determines human blood groups. Named for the presence of A and B carbohydrates on the surface of the cell, or the absence of the carbohydrates in the case of the O group. This system uses four possible combinations: A, B, AB, or O.

Acetylcholine (ACh) A neurotransmitter released in the central and peripheral nervous system, specifically at neuromuscular joints.

Acetyl Coenzyme A (acetyl CoA) A molecule that enters the citric acid cycle to produce energy. The acetyl CoA can come from sugars that have gone through glycolysis, or it can come directly from fats or proteins in the cell.

Acidosis An abnormal increase in the acidity of the body's fluids.

Acquired Immunity A type of immunity that is not the result of genetic inheritance, but rather due to the exposure to some antigen and the resulting response by the immune system.

Actin One of the major proteins involved in muscle contraction. Actin proteins form a long fiber within the muscle contractile unit. Myosin, the other major protein involved in muscle contraction, attaches to the actin filaments and pulls the muscle shorter.

Action Potential A change in the electrical charge of a nerve cell following the transmission of a nerve impulse.

Adaptation The state of sensory acclimation in which the sensory awareness diminishes despite the continuation of the stimulus.

Adduction Movement of a limb toward the median line of the body.

Adenosine Diphosphate (ADP) A chemical substance produced through digestion and used in cell respiration and energy production.

Adenosine Triphosphate (ATP) A chemical substance produced from aerobic cell respiration that is the muscle's direct source of energy for movement.

Adipocytes Cells that have large holes filled with fat.

Adipose Capsule The central layer surrounding the kidney, composed of fatty tissue.

Adrenal Glands The hormone-releasing glands located above the kidneys.

Adrenaline Also known as epinephrine, a hormone produced by the adrenal glands that helps regulate the sympathetic division of the autonomic nervous system. During times of stress or fear, the body produces additional amounts of adrenaline into the bloodstream, causing an increase in blood pressure and cardiac activity.

Adrenocorticotropic Hormone (ACTH) Hormone produced by the pituitary gland, which stimulates the release of hormones from the adrenal cortex. Also called corticotropin.

Aerobic Anything having to do with acquiring oxygen from the air.

Aerobic Cell Respiration A chemical process that allows the cell to produce energy from glucose and oxygen.

Aerobic Exercise Exercise in which energy is made by processes involving oxygen. Types of aerobic exercise include swimming, biking, and jogging.

Afferent Nerves Fibers coming to the central nervous system from the muscles, joints, skin, or internal organs.

Afferent Vessels A form of vessel that brings fluid towards an organ or lymph node.

After-image An image of a visual nature that exists even after the visual stimulus has ceased.

Agglutination The clumping of blood. This can occur if a patient with a certain blood type is given blood of another type.

Agonists Hormones that bind to their receptor and elicit a specific biological response.

Albumin The most abundant plasma protein. It makes up 55 percent of the total protein content of plasma. It is involved in maintaining blood volume and water concentration.

Aldosterone A hormone secreted by the adrenal glands in the kidneys that increases sodium reabsorption.

Alkaline A term used to indicate a pH of 7 or greater. Sometimes also called basic.

Allele A variation of a gene that encodes for a specific trait. It is usually due to minor variations in the DNA at the molecular level.

Allergies Hypersensitive reaction to a particular substance or allergen; symptoms vary in intensity.

Alveoli Tiny air sacs in the lungs. They exchange oxygen and carbon dioxide between the lungs and the blood.

Amino Acids The building blocks of proteins.

Amphiarthrosis Joint that permits only slight movement.

Amphiphatic Molecules A term given to a molecule that has both hydrophilic and hydrophobic properties.

Anaerobic Exercise Exercise in which energy is made by processes that do not involve oxygen. Types of anaerobic exercise include weight lifting and sprinting.

Androgens Male sex hormones produced by the gonads and adrenal cortex.

Anemia A reduction in the number of red blood cells in the body, resulting in an insufficient number of hemoglobin molecules to carry oxygen to the tissues of the body. This may result in tissue color changes, weakness, and increased susceptibility to disease.

Anions Negatively charged particles.

Annulus Fibrosus A ring of fibrous connective tissue that serves as an anchor for the heart muscle and as an almost continuous electrical barrier between the atria and ventricles.

Antagonistic Muscle Pair Two muscles that have an opposite action, such as the muscle that bends the arm and the muscle that straightens the arm. The antagonistic pair controls and stabilizes the elbow as it bends and straightens.

Antagonists Hormones that bind to the receptor but do not trigger a biological response. By occupying the receptor, an antagonist blocks an agonist from binding and thus prevents the triggering of the desired effect within the cell.

Anterior Situated in front; at or toward the head end of a person or animal.

Anterior Pituitary The lobe of the pituitary that secretes hormones that stimulate the adrenal glands, thyroid gland, ovaries, and testes.

Antibody Proteins that attack antigens.

Antidiuretic Hormone (ADH) Hormone produced by the pituitary gland that increases the permeability of the kidney ducts to return more fluid to the bloodstream. Also called vasopressin.

Antigens Invading organisms and materials that enter the human body. The body may mount a defense with antibodies.

Antioxidants Compounds that prevent oxidative damage to organic molecules. Vitamins C and E are examples of antioxidant nutrients, as is the mineral selenium.

Aorta The largest artery in the human body. It supplies oxygenated blood from the left ventricle of the heart to the branching arteries, which in turn supply oxygen to all parts of the body.

Aortic Arch A large, rounded section of the aorta that occurs above the heart, just after the aorta leaves the right ventricle.

Aortic Bodies Chemoreceptors found in the aortic arch, the curved portion between the ascending and descending parts of the aorta.

Appendectomy The surgical procedure that is used to remove an inflamed,
diseased, or ruptured appendix.

Appendicitis An inflammation of the appendix. This is usually caused by an infection of the appendix and results in fever, pain, and loss of appetite.

Appendicular Skeleton The skeletal structures composing and supporting the appendages; these include the bones of the shoulder and hip girdles as well as those of the arms and legs.

Arterial Baroreceptor Reflex The mechanism that provides oversight and maintenance of the blood flow by responding to slight changes in blood pressure.

Arterial System The portion of the circulatory system that delivers oxygen-rich blood to the body tissues.

Arteries Larger blood vessels that deliver oxygen-rich blood to the body tissues.

Arterioles Smaller blood vessels that deliver oxygen-rich blood to the body tissues.

Arteriovenous Anastomoses Blood vessels that directly connect arterioles to venules. Commonly, blood travels from arterioles to capillaries to venules. Arteriovenous anastomoses are typically found in only a few tissues.

Arthritis An inflammation of the joints.

Asexual Reproduction Reproduction in which genetically identical offspring are produced from a single parent.

Atherosclerosis Also known as hardening of the arteries. It is a narrowing of arterial walls caused by deposits, collectively called plaque, that create rough, irregular surfaces prone to blood clots.

Atria Plural of atrium.

Atrium In the human heart, it is one of the heart's two upper chambers. The plural form is atria.

Autocatalytic Process A chemical reaction in which the products of the reaction are responsible for initiating the start of the reaction.

Autocrine The action of a hormone on the cells that produced it.

Autoimmune A term used to describe an immune response to the patient's own body. An autoimmune disease is therefore one that attacks part of the patient's body.

Autoimmune Disease This occurs when the immune system incorrectly identifies the tissues of the body as foreign material, and begins an immune response against the cells or tissue. Lupus and forms of diabetes may be caused by an autoimmune response.

Autonomic Nervous System The part of the nervous system that controls involuntary actions and rules the variations of the heart rate.

Axial Skeleton The central supporting portion of the skeleton, composed of the skull, vertebral column, ribs, and breastbone.

Axon A single nerve fiber that carries impulses away from the cell body and the dendrites.

B Cells Also known as B lymphocytes. They are one of two main types of lymphocyte, and participate in the body's immune response.

B Lymphocytes *See* B Cells.

Baroreceptors Pressure detectors located in the major arteries. Part of the arterial baroreceptor reflex, they sense a dip or spike in blood pressure.

Basilar Artery A blood vessel that arises from the vertebral arteries and joins with other cerebral arteries to form the circle of Willis.

Basophil A type of granulocyte that appears to be active in the inflammatory process.

Bayliss Myogenic Response The mechanism by which smooth muscle cells impart muscle tone to the blood vessels.

Bilirubin A waste product produced by the liver that is the result of the breakdown of red blood cells. It is released into the small intestine, but some is reabsorbed back into the blood and excreted with the urine.

Binucleate Cell A cell that contains two nuclei.

Bioavailability A term of nutritional analysis that indicates how much of a nutrient in a food is actually available to the body for absorption by the gastrointestinal tract.

Biomarker A molecular clue indicating the presence of disease or the genetic predisposition for disease.

Biomolecule A general classification for any of the four groups of organic molecules that are used in the building of cells—proteins, carbohydrates, lipids, and nucleic acids.

Bipedal An animal that walks on two feet.

Bladder A hollow, muscular organ that stores urine for elimination.

Blastocyst An embryonic stage following the morula stage characterized by outer trophoblast cells, an inner cell mass, and a central, fluid-filled cavity.

Blood The fluid that contains the plasma, blood cells, and proteins and carries oxygen, carbon dioxide, nutrients, waste products, and other molecules throughout the body.

Blood Cells Cells contained in the plasma of the blood. *See also* Red Blood Cells and White Blood Cells.

Blood Pressure The force of the blood against the walls of the blood vessels.

Blood Sugar Level The amount of glucose in the blood.

Blood Type A form of blood, determined by the presence or absence of chemical molecules on red blood cells. A person may have type A, B, O, or AB blood.

Blood Vessels Also known as the vasculature. These are the tubes of the circulatory system that transport the blood throughout the body.

Bohr Effect High concentrations of carbon dioxide and hydrogen ions in the capillaries in metabolically active tissue that decrease the affinity of hemoglobin for oxygen and leads to a shift to the right in the oxygen dissociation curve.

Bolus The name given to the mass of food that accumulates at the rear of the oral cavity for swallowing.

Bone Marrow The site in the body where the cells of the lymphatic system originate.

Bowman's Capsule The bulb surrounding the glomeruli. It provides an efficient transfer site for water and waste products to move from the blood to the urinary system.

Brachial Artery The blood vessel in the upper arm that accepts blood from the subclavian artery by way of the axillary artery, and travels down the arm to supply the ulnar, radial, and other arteries of the forearm.

Brachial Vein The blood vessel that collects blood from the ulnar vein and empties into the axillary vein.

Brachiocephalic Artery Also known as the innominate artery. This short blood vessel arises from the aortic arch, and branches into the right common carotid artery and the right subclavian artery.

Brachiocephalic Veins Also known as the innominate veins. This pair of veins arises from the convergence of the internal jugular and subclavian veins and flows into the superior vena cava.

Brain Other than the spinal cord, the primary organ in the nervous system.

Brain Stem This area of the brain connects the cerebrum with the spinal cord and is also the general term for the area between the thalamus and the spinal cord, which includes the medulla and pons.

Bronchi The two large air tubes leading from the trachea to the lungs that convey air to and from the lungs.

Bronchial Vein One of two main blood vessels that collect newly oxygenated blood from the bronchi and a portion of the lungs, and deliver it through one or more smaller veins to the superior vena cava.

Bronchiole Any of the smallest bronchial tubes that end in alveoli.

Buccal Cavity Another term commonly used to describe the oral cavity. It technically represents the space between the back of the teeth and gums to the rear of the mouth.

Bundle of His A thick, conductive tract located in the heart that transmits the electrical signal from the AV node to the Purkinje fibers in the base of the ventricle wall.

Bursa A sac of fluid within a joint.

Calcitonin Hormone produced by the thyroid gland that influences calcium and phosphorous levels in the blood.

Calcitrol A hormone secreted by the kidneys that increases the levels of calcium and phosphorous in the blood.

Calorie A measure of how much energy food contains.

Cancellous Bone Bone that has a latticework structure, such as the spongy tissue in the trabecular bone.

Capillaries The tiniest blood vessels. They are the sites of exchange: At body tissues, blood in the capillaries delivers oxygen and nutrients, and picks up carbon dioxide and waste products; and at the lungs, blood in the capillaries drops off carbon dioxide and picks up oxygen.

Carbon Monoxide Poisoning A medical condition arising when a person is exposed to carbon monoxide gas. Prolonged exposure can be fatal.

Cardiac (Heart) Muscle The type of muscle found in the heart.

Catecholamines A class of hormone (including epinephrine and norepinephrine) synthesized in the adrenal medulla that is involved in the body's stress response.

Cation A positively charged particle.

Cell Body Main mass of the neuron that contains the nucleus and organelles.

Central Nervous System Division of the nervous system that contains the brain and the spinal cord.

Cerebellum Located towards the back of the medulla and pons, this portion of the brain is in charge of many subconscious aspects of skeletal muscle functioning, such as coordination and muscle tone.

Cerebral Aqueduct The tunnel that runs through the midbrain, allowing cerebrospinal fluid to travel from the third to the fourth ventricle.

Cerebral Cortex This area of the brain is the gray matter located on the surface of the cerebral hemispheres. The cerebral cortex includes the brain's motor, sensory, auditory, visual, taste, olfactory, speech, and association areas.

Cerebrospinal Fluid (CSF) The fluid in the spinal cord's central canal that serves as the fluid for the central nervous system. This tissue fluid circulates in and around the brain.

Cerebrum This is the largest portion of the brain and consists of the left and right cerebral hemispheres. The cerebrum controls movement, sensation, learning, and memory.

Chemoreceptors Cells that respond to changes in their chemical environment by creating nerve impulses. Some chemoreceptors in the brain respond to carbon dioxide levels in the blood to help regulate breathing.

Chemotaxis The reaction of mobile cells to a chemical gradient; the cells may move either towards or away from the gradient depending on the nature of the chemical being used.

Chloride Shift Describes the exchange of negatively charged chloride ions for negatively charged bicarbonate ions across an erythrocyte's cell membrane.

Cholesterol A fatlike substance that occurs naturally in the body. Two types exist: high-density lipoprotein (HDL) and low-density lipoprotein (LDL).

Chondrocytes Cartilage cells.

Chordae Tendineae Tiny tendinous cords located at each of the heart valves. They attach to nearby muscles and prevent blood backflow through the valves.

Choroid Plexus This capillary network helps form the cerebrospinal fluid in the brain.

Chromatin A diffuse mixture of DNA and proteins that condenses into chromosomes prior to cell division.

Chromosomes Cellular structures composed of proteins and DNA that carry the body's hereditary information.

Cilia Hairlike projections from the surface of a cell. In the respiratory system, cilia help filter out foreign particles from the air before they reach the lungs.

Circadian Rhythm The body's 24-hour biological cycle that regulates certain activities, such as sleep, regardless of environmental conditions, including lightness and darkness.

Circle of Willis A vascular structure that supplies blood to the brain. It arises from the basilar, internal carotid, and other arteries.

Circulatory System The heart, blood vessels, and blood.

Circumcision The surgical removal of the foreskin. The term is also sometimes used with reference to females to describe a controversial and excruciating practice of genital mutilation that is common in certain societies around the world.

Citric Acid Cycle A chemical reaction that takes place in the mitochondria. The cycle produces some energy for the cell and produces products that can be used to produce large amounts of energy through oxidative phosphorylation.

Colic Arteries Divided into right, left, and middle colic arteries, all of which branch from either the inferior or superior mesenteric arteries, and feed the colon.

Collagen The albumin-like substance in connective tissue, cartilage, and bone.

Collecting Duct Where fluid is carried from the distal convoluted tubule (DCT) in the nephron of the kidneys on its way to the minor calyx.

Colostrum Nutritious fluids secreted by the breasts shortly before and after a woman gives birth; precedes the production of breast milk.

Common Carotid Arteries One of two major blood vessels that supply the head. The left carotid splits directly from the aortic arch between the bases of the two coronary arteries. The right carotid indirectly branches from the aorta via the brachiocephalic artery.

Complement The collective term for a variety of beta globulins. *See also* Globulins.

Complement Fixation The process by which complement factors bind to either antibodies or cell surfaces during the immune response.

Complementary Base Pair Nucleotide bases (adenine and thymine or guanine and cytosine) that pair up via hydrogen bonds in DNA.

Compression Forces Forces that squeeze items together; blows that press against the body.

Computed Tomography (CT) Scan A commonly used tool for determining the nature of a stroke.

Concentration Gradient The change in solute concentration from one location to another. Unless restricted, solutes will move from a site of higher solute
concentration to one of lower solute concentration, leading to an equilibrium between the two sites.

Concentric Contraction The type of contraction that occurs when a muscle contracts and grows shorter, such as the biceps muscle when bending the elbow.

Conchae Structures or parts that resemble a seashell in shape with three bony ridges or projections—the superior, middle, and inferior conchae—on the surface of the nasal cavity sides.

Contraceptive An agent that prevents ovulation, kills sperm, or blocks sperm from reaching the ovum for fertilization.

Convergence An impulse pathway where a neuron receives impulses from the nerve endings of thousands of other neurons but transmits its message to only a few other neurons.

Coronal Plane Divides the body into front and back portions.

Coronary Arteries Arising from the base of the aorta, these are the two major arteries that feed the heart muscle. The right coronary artery remains a single, large vessel, but the left coronary artery almost immediately splits into transverse and descending branches.

Coronary Circulation The circulatory system of the heart.

Corpus callosum A band of white matter connecting the cerebral hemispheres.

Corpus luteum Progesterone-secreting tissue that forms from a ruptured Graafian follicle in the mammalian ovary after the egg has been released.

Cortex The tissue layer that covers the brain.

Cortical Bone The hard, dense bone that forms the outer shell of all bones.

Corticotroph Cell in the anterior pituitary gland that secretes corticotropin (ACTH).

Cortisol A steroid hormone produced by the adrenal cortex that influences the body's stress response.

Cranial Nerves The brain's 12 pairs of nerves located in the peripheral nervous system.

Craniosacral Division Another name for the parasympathetic division of the autonomic nervous system. In this division, all the cell bodies of pre-ganglionic neurons are located in the brain stem and sacral segments of the spinal cord.

Cranium The bones of the skull that house the brain.

Creatinine Waste produced by the breakdown of creatine phosphate in muscles.

Creatine Phosphate A molecule stored in the muscle that can quickly replenish ATP during a sudden burst of exercise.

CT Scan *See* Computed Tomography (CT) Scan.

Cuboid Bones Bones in the wrist that are shaped like cubes.

Cutaneous Senses The skin's sensory mechanisms whose receptors are located in the dermis.

Cytochromes A class of membrane-bound intracellular hemoprotein respiratory pigments. These enzymes function in electron transport as carriers of electrons.

Cytokines Signaling peptides secreted by immune cells and other types of cells in response to infection or other stimuli.

Cytoplasm Cellular material located between the nucleus and cell membrane.

Daughter Cells Cells arising from mitotic division that are identical to the parent cell.

Deglutition Another term used for the act of swallowing.

Dehydration Synthesis A form of chemical reaction that involves the removal of water to form a chemical bond. Also called a condensation reaction.

Dentin A tissue that is the majority of the mass of a tooth. It consists primarily of minerals (70 percent), with the remainder being water and organic material.

Depolarization When the electrical charges in a nerve cell reverse due to a stimulus. The rapid infusion of sodium ions causes a negative charge outside and a positive charge inside the cell membrane.

Detrusor Muscle Three layers of smooth muscle surrounding the mucosa of the bladder.

Diaphragm A muscle that aids in respiration. It separates the thoracic cavity from the abdominal cavity.

Diaphysis The central shaft of a bone.

Diastole The heart's resting period.

Diathroses Joints allowing free movement.

Diffusion The passive flow of molecules from one location to another.

Diploid Cells Any organism whose cells contain two copies of each chromosome. The majority of human cells, except sex cells and some liver cells, are diploid.

Dislocation Condition when a bone is moved out of a joint.

Distal Indicates direction away from the torso.

Distal Convoluted Tubule (DCT) Located between the loop of Henle and the collecting duct inside the nephron of the kidney.

Diuretic A substance (i.e., caffeine) that increases urine production.

Divergence An impulse pathway where a neuron receives impulses from a few other neurons and relays these impulses to thousands of other neurons.

DNA A nucleic acid that contains a cell's genetic or hereditary information.

Dopamine A neurotransmitter found in the motor system, limbic system, and the hypothalamus.

Dorsal Root The sensory root of a spinal nerve that attaches the nerve to the posterior part of the spinal cord.

Dorsal Root Ganglion An enlarged portion of the spinal nerve's dorsal root that contains the sensory neuron's cell bodies.

Down Syndrome Mental retardation associated with specific chromosomal abnormalities.

Dura Mater This fibrous connective tissue is the outermost layer of the brain's meninges.

Eccentric Contraction The type of contraction that occurs when a muscle contracts but the overall muscle grows longer rather than shorter; an eccentric contraction occurs in the biceps which contracts to control the arm as it extends, but the muscle grows longer rather than shorter.

Effector A muscle, gland, or other organ that responds after receiving an impulse.

Efferent Nerves Fibers leaving the central nervous system carrying messages to the muscles, joints, skin, or internal organs.

Efferent Neuron Nerve cells that carry impulses and messages away from the spinal cord and brain to the muscles and glands.

Efferent Vessels A vessel of the lymphatic or circulatory systems that carries fluid away from an organ or lymph node.

Eicosanoids Compounds derived from fatty acids that act like hormones to influence physiologic functions.

Elastin A protein of blood vessels that imparts elasticity.

Electrocardiogram (ECG or EKG) The product of an electrocardiograph, it is a printout depicting the heart's electrical activity. An ECG has five parts, each signified with the letter P, Q, R, S, or T, that reflect different phases in the heart activity.

Electrocardiograph (ECG or EKG) A device that records the heart's electrical activity as a jagged line on a sheet of paper, which is called an electrocardiogram.

Electrolyte A charged particle like calcium (Ca^{2+}) or magnesium (Mg^{2+}) that may have a number of functions in cells.

Electron Transport System A complex sequence found in the mitochondrial membrane that accepts electrons from electron donors and then passes them across the mitochondrial membrane creating an electrical and chemical gradient.

Embryogenesis The entire process of cell division and differentiation leading to the formation of an embryo.

End-diastolic Volume The amount of blood in a completely filled ventricle. In an adult, this is typically about 0.12 quarts (120 ml).

Endocardium The membrane lining the heart.

Endocrine System The body's organ system that controls hormone secretion.

Endothelium In blood vessels, it is also known as the tunica intima. The tunica intima forms the innermost layer of blood vessels.

Eosinophil A type of granulocyte that appears to be active in the moderation of allergic responses and the destruction of parasites.

Epiblast The outer layer of a blastocyst before differentiation into the ectoderm, mesoderm, or endoderm.

Epidemic A widespread outbreak of an infectious disease that affects a disproportionately large number of people within a given population.

Epinephrine *See* Adrenaline.

Epiphysis The portion of bone attached to another bone by a layer of cartilage.

Epithelial Cell A type of cell that lines organs and tissues of the body. It specializes in the exchange of materials with the external environment, such as the lumen of the gastrointestinal tract.

Epitope The specific area of an antigen to which the B cell receptor binds.

Equilibrium Balance mechanisms that are regulated by inner ear structures.

Erythroblast An early stage in red blood cell development.

Erythrocytes Red blood cells.

Erythropoietin A protein hormone produced by the kidneys that stimulates red blood cell production.

Estrogen Any of a family of hormones produced by the female ovaries that determine female sexual characteristics and influence reproductive development.

Excitatory Nerve/Fiber A nerve fiber that passes impulses on to other fibers.

Excitatory Synapse The passing of an impulse transmission to other synapses.

Exocrine Glands Glands that utilize ducts to release their secretions to the outside environment.

Extension A stretching out, as in straightening a limb.

External Urethral Sphincter Ring of voluntary muscle surrounding the end of the urethra, which regulates urine flow out of the body.

Extracellular Fluid The water found outside a cell that contains plasma and other tissue fluids.

Extrinsic Factors Another term used for vitamin B_{12} in the diet.

Facilitated Diffusion A passive process that utilizes a membrane-bound protein to move a compound across a membrane down its concentration gradient.

Fascia The connective tissue surrounding an entire muscle. The fascia becomes part of the tendon at either end of the muscle, connecting the muscle to the bone.

Fascicle A bundle of muscle fibers within the muscle surrounded by a tissue called the perimysium. Each muscle is made up of many fascicles.

Fast-twitch Muscle A type of muscle fiber that is able to contract very quickly. These fibers are predominantly found in muscles that must contract quickly and with great strength but do not need to contract over a long period of time.

Femoral Artery Arising from one of the two external iliac arteries, the femoral artery traverses the thigh to the popliteal artery.

Femoral Vein A large blood vessel in the thigh that collects blood from the popliteal vein and great saphenous vein and delivers it to the external iliac vein.

Fibrinogen A protein in plasma. It functions in blood clotting.

Fibular Vein *See* Peroneal Veins.

Flavoproteins The enzymes that contain flavin bound to a protein. Flavoproteins play a major role in biological oxidations.

Flexion The bending of a joint or of body parts having joints.

Follicle-stimulating Hormone (FSH) A hormone produced by the anterior pituitary gland that triggers sperm production in the testes and stimulates the development of follicles in the ovaries.

Fossa A hole or indentation.

Fossae The plural form of fossa.

Gametes Reproductive cells that, before fusing at fertilization, are haploid—they contain 23 instead of 46 chromosomes.

Gas Exchange In the respiratory system, gas exchange refers to the process of acquiring oxygen from the air and eliminating carbon dioxide from the blood.

Gastric Arteries Blood vessels of the digestive system. The left gastric artery stems from the celiac artery and supplies the stomach and lower part of the esophagus. The right gastric artery stems from the common hepatic artery and eventually connects with the left gastric artery.

Gastric Inhibitory Peptide (GIP) The gastrointestinal hormone whose main action is to block the secretion of gastric acid.

Gastric Veins Blood vessels of the digestive system. Blood from the stomach exits into the gastric veins, which then empty into a number of other veins that ultimately enter the portal vein (in the case of the left and right gastric veins) or the splenic vein (in the case of the short gastric vein).

Gastrin Hormone produced by the gastrointestinal system that regulates stomach acid secretion.

Gastroepiploic Arteries Blood vessels of the digestive system. The right gastroepiploic artery branches from the gastroduodenal artery. The left gastroepiploic artery branches from the splenic artery. Both provide blood to the stomach and duodenum.

Gene Expression In genetics, a term describing the results of activating of a gene.

Gene Transcription The process by which a strand of DNA is copied to form a complementary RNA strand.

Genetic Immunity A form of immunity to a pathogen that is inherited.

Genetic Imprinting Refers to differences in the way maternal or paternal genes are expressed in the offspring.

Genetic Sex Gender determination based on an XX or an XY chromosome configuration.

Glia Support cells in the brain.

Globulins Plasma proteins that function as transportation vehicles for a variety of molecules, in blood clotting and/or in the body's immune responses. They are divided into three types alpha, beta, and gamma globulins.

Glomerular Capsule A cup-shaped sac that surrounds glomeruli of the nephrons.

Glomerular Filtrate The product of blood filtration in the nephrons of the kidneys.

Glomeruli Clusters of capillaries in the kidneys.

Glomerulus The singular form of glomeruli.

Glossopharyngeal Nerve The mixed nerve in the throat and salivary glands that contains sensory fibers for the throat and taste from the posterior one-third of the tongue.

Glucose A form of sugar that is a necessary component (along with oxygen) in cell respiration.

Glucose Tolerance Test A test measuring blood sugar levels that is often used to diagnose diabetes.

Glutamate A neurotransmitter associated with pain-related impulses.

Glycogen A storage form of carbohydrate. The liver converts fats, amino acids, and sugars to glycogen, which functions as a reserve energy supply for the body.

Glycogenolysis The breakdown of glycogen in the liver and in muscle tissue.

Glycolysis The process of breaking down glucose into two molecules of pyruvate. Glycolysis produces some energy for the cell and is the primary way of producing energy during anaerobic exercise.

Glycoprotein An organic compound composed of a joined protein and carbohydrate.

Goblet Cell An epithelial cell that secretes mucus.

Gonadotroph A cell in the anterior pituitary gland that secretes luteinizing hormone and follicle-stimulating hormone.

Gonadotropins Hormones (luteinizing hormone and follicle-stimulating hormone) released by the anterior pituitary gland that stimulate the ovaries and testes.

Graft Rejection The tendency of the immune system to reject transplanted tissue as foreign.

Granulocyte The most abundant type of white blood cell. *See also* Basophil, Eosinophil, and Neutrophil.

Gray Matter Nerve tissue located in the central nervous system containing cell bodies of neurons.

Growth Factors Proteins that act on cells to stimulate differentiation and proliferation.

Growth Hormone Hormone secreted by the anterior pituitary gland that promotes bone and muscle growth and metabolism.

Growth Hormone-Releasing Hormone (GHRH) A hormone that stimulates the anterior pituitary gland to secrete the growth hormone (GH).

Gyri Folds or ridges in the cerebral cortex.

H Zone The space between the two sets of actin filaments in the center of the sarcomere. The H zone grows smaller when the sarcomere contracts, and the actin filaments slide toward each other in the center of the sarcomere.

Haldane Effect A high concentration of oxygen, such as occurs in the alveolar capillaries of the lungs, that promotes the dissociation of carbon dioxide and hydrogen ions from hemoglobin.

HDL *See* High-density Lipoprotein (HDL).

Heart The muscular pump that powers the circulatory system.

Heart Attack Also known as a myocardial infarction, this condition happens when the supply of oxygen to a portion of the heart muscle is curtailed to such a degree that the tissue dies or sustains permanent damage.

Heart Failure A condition in which the heart can no longer carry out its pumping function adequately, resulting in slow blood circulation, poorly oxygenated cells, and veins that hold more blood.

Hematopoeisis The formation and maturation of blood cells.

Hematuria Blood in the urine.

Heme Group A ringlike chemical structure that is part of hemoglobin.

Hemes The deep-red organic pigment that contains iron and other atoms to which oxygen binds in blood hemoglobin. Hemes are found in most oxygen-carrying proteins.

Hemoglobin A large chemical compound in red blood cells that imparts their red color and also participates in transporting oxygen and carbon dioxide.

Hemolysis The rupture and destruction of red blood cells.

Hepatic Arteries The blood vessels supplying the liver and other organs. The common hepatic artery arises from the celiac trunk and supplies the right gastric, gastroduodenal, and proper hepatic arteries. The proper hepatic artery supplies the liver by way of the cystic artery.

Hepatic Portal System The name given to the portion of the circulatory system that connects the stomach and both intestines to the liver.

Hepatic Vein The blood vessel that collects blood from the liver and delivers it to the inferior vena cava.

High Blood Pressure A medical condition that arises when the pressure of the blood against the blood vessel walls exceeds normal limits. It results from a narrowing of the arterioles.

High-Density Lipoprotein (HDL) Often called the type of cholesterol curtails the accumulation of low-density lipoprotein in blood vessels.

Hilius The curved notch on the side of each kidney near the center where blood vessels enter and exit the kidney.

Histones Proteins associated with gene expression.

Homeostasis The regulation of the body's internal environment to maintain balance.

Homologous In genetics, chromosomes (one from the male parent, one from the female parent) carrying alleles for similar traits, such as eye color, that pair up during meiosis.

Hormone A chemical compound, often called a chemical messenger, that the brain and other organs use to communicate with the cells.

Huntington's Chorea A progressive and fatal disease affecting the nervous system.

Hydrolysis In chemistry, the breaking of a chemical bond by the addition of water.

Hydrophilic A water-loving compound, meaning that it is soluble in water. An example is glucose.

Hydrophobic A water-fearing compound, meaning that is it generally insoluble in water. Most lipids are hydrophobic, as are some amino acids.

Hypernatremia Too much sodium in the extracellular fluid.

Hypertension *See* High Blood Pressure.

Hypertrophy The process in which muscles grow larger in response to exercise.

Hyperventilation An increased and excessive depth and rate of breathing greater than demanded by the body's needs; can lead to abnormal loss of carbon dioxide from the blood, dizziness, tingling of the fingers and toes, and chest pain.

Hypoblast The inner layer of tissue in a developing embryo that will eventually become the digestive tract and respiratory tract.

Hypocalcemia A deficiency of calcium in the blood.

Hyponatremia Too little sodium in the extracellular fluid.

Hypophysis The pituitary gland.

Hypothalamic-Hypophyseal Portal System The circulation system through which neurohormones from the hypothalamus travel directly to the anterior pituitary gland without ever entering the general circulation.

Hypothalamic-Pituitary-Target Organ Axis A multiloop feedback system that coordinates the efforts of the hypothalamus, the pituitary gland, and the target gland.

Hypothalamus This part of the brain regulates body temperature and pituitary gland secretions. The hypothalamus is located superior to the pituitary gland and inferior to the thalamus.

Hypoxia A sudden decrease in the blood's oxygen content.

I Band The region between the Z band at the outside of the sarcomere and the end of the myosin chain that spans the center of the sarcomere.

Iliac Arteries These arise at the end of the abdominal artery. The abdominal artery bifurcates into two common iliac arteries, each of which soon divides again into internal and external iliac arteries.

Iliac Veins Blood from the femoral vein collects in the external iliac vein, which joins the internal iliac vein and carries blood from the pelvis to form the common iliac vein.

Immune System A body system that includes the thymus and bone marrow and lymphoid tissues. The immune system protects the body from foreign substances and pathogenic organisms in the form of specialized cellular responses.

Immunity The ability of an organism not to be affected by a given disease or pathogen.

Immunoglobulins (Ig) Plasma proteins that act as antibodies. The five main types are IgA, IgD, IgE, IgG, and IgM.

Inflammatory Mediators Soluble, diffusible molecules that act locally at the site of tissue damage and infection.

Inhibin Hormone secreted by the ovaries and testes that inhibits the release of follicle-stimulating hormone (FSH) by the pituitary.

Inhibitory Nerve A type of nerve fiber that obstructs impulse transmission to another fiber.

Inhibitory Synapse An impulse transmission obstruction due to a chemical inactivator located at the dendrite of the postsynaptic neuron.

In-series Blood Circulation Also known as portal circulation. It is blood flow that travels from one organ to another in series.

Insertion The end of the muscle that is usually farthest from the center of the body and usually the one that moves when the muscle contracts.

Insulin A hormone secreted by the pancreas. It allows the body cells to use energy, specifically glucose.

Insulin-like Growth Factors Substances produced in the liver and other tissues that act much like growth hormone, stimulating bone, cartilage, and muscle cell growth and differentiation.

Intercalated Disk A disk that separates two muscle fibers in the heart muscle. This disk can conduct the signal to contract from one muscle fiber to the next. With this connection, the entire heart muscle can contract in unison.

Intercostal Muscles Found under the ribs, these muscles play a role in respiration.

Interferons A family of drugs used to regulate the body's immune system. They may be used for such diseases as multiple sclerosis or cirrhosis of the liver.

Interlobar Arteries Blood vessels that branch from the renal artery to disperse blood throughout the kidney and to glomeruli.

Intermediate Pituitary A lobe of the pituitary of which only vestiges remain in humans.

Intermediolateral Cell Column Located on the thoracic level of the spinal cord, this is an extra cell column where all presynaptic sympathetic nerve cell bodies are located.

Internal Urethral Sphincter Ring of involuntary muscle that surrounds the urethra where it meets the bladder and that controls the flow of urine.

Intestinal Villi Tiny projections that line the inside wall of the small intestine and the uptake of nutrients by capillaries.

Intracapsular Ligaments Ligaments within the capsule at the joint.

Intracellular Fluid The water found within a cell.

Intrinsic Factors A protein released by the gastrointestinal tract that aids in the absorption of vitamin B_{12}.

In Vitro Occurring outside the body, often used to refer to laboratory procedures such as fertilization of ova within a laboratory dish.

In Vivo Occurring inside the body.

Involuntary Muscle *See* Smooth Muscle.

Ions Any element or compound that loses or gains electrons and in the process changes its net electric charge.

Islets of Langerhans Endocrine cells located in the pancreas in which the hormones insulin and glucagon are produced.

Isometric Contraction The type of contraction that occurs when a muscle contracts but the joint does not open or close, such as when pushing against a wall or pushing down on a table.

Joint The union between two bones.

Jugular Veins Blood vessels of the head and/or neck. The anterior jugular vein collects blood from veins of the lower face, traverses the front of the neck, and delivers the blood to the external jugular vein. The external jugular vein is a large vein that also receives blood from within the face and around the outside of the cranium, and empties into one of several veins, including the internal jugular. The internal jugular vein is the largest vein of the head and neck, and also drains blood from the brain and neck. It joins the subclavian vein to form the brachiocephalic vein.

Ketone Bodies Substances produced from fats when not enough glucose is present, which provide an alternate energy source for the brain and other tissues.

Kidneys The two bean-shaped organs that filter wastes, regulate electrolyte balance, and secrete hormones.

Kilocalorie The amount of energy required to raise 1,000 grams of water from 14.5° to 15.5° Celsius at standard atmospheric pressure.

Lacteals The portion of the lymphatic system that is associated with the gastrointestinal system, specifically the intestines.

Lactotroph A cell in the anterior pituitary gland that secretes prolactin.

Lateral Situated on a side.

LDL *See* Low-density Lipoprotein (LDL).

Leptin A protein hormone that influences metabolism and regulates body fat.

Ligament A tough band of connective tissue that connects bones to each other.

Lipoproteins Proteins that are connected chemically to lipids and used by the digestive system to transport hydrophobic fats and lipids in the hydrophilic bloodstream.

Loop of Henle The U-shaped section between the proximal convoluted tubule and the distal convoluted tubule in the nephron of the kidney.

Low-density Lipoprotein (LDL) Often called the "bad" cholesterol. This type of cholesterol can build up on blood-vessel walls and cause health problems.

Lumbar Veins Blood vessels of the digestive system. Lumbar veins collect blood from the abdominal walls and deliver it to other veins, including the inferior vena cava.

Lumen The internal diameter of a blood vessel. It represents the open space in the vessel through which the blood flows.

Luteinizing Hormone (LH) A hormone produced and secreted by the anterior pituitary gland that stimulates ovulation and menstruation in women and androgen synthesis by the testes in men.

Luteolysis The process by which the corpus luteum in the ovary degenerates when an egg is not fertilized.

Lymph Fluid in the vessels of the lymphatic system. It is the interstitial fluid that exits the capillaries and enters surrounding cells during the capillaries' exchange function.

Lymphatic System A series of vessels that shunts excess tissue fluid into the veins.

Lymph Node Filters that separate from lymph any invading organisms and other foreign materials.

Lymphocyte A type of leucocyte that detects antigens and serves in the body's immune response. The two main types are B cells and T cells.

Macrophage White blood cells that ingest and digest bacteria, other foreign organisms, platelets, and old or deformed red blood cells.

Magnetic Resonance Imaging (MRI) A diagnostic tool for viewing blood flow and locating sites of blood-flow blockage.

Major Calyx Openings in the center of the kidneys through which urine flows into the renal pelvis.

Malignant A condition that becomes progressively worse or more pronounced over time, and which may lead to death.

Medial Toward the midline of the body.

Medulla Located above the spinal cord, this part of the brain controls vital functions such as heart rate, respiration, and blood pressure.

Medullary Cords Within a lymph node, these are areas of dense lymphatic tissue.

Meiosis A process of cell division resulting in daughter cells containing half the number of chromosomes contained in the parent cell. In humans, this process is responsible for the generation of the sex cells, oocytes and sperm.

Meninges The membrane is composed of connective tissue that covers the brain and spinal cord and lines the dorsal cavity.

Mesenteric Arteries Blood vessels of the digestive system. The inferior and superior mesenteric arteries arise from the abdominal aorta and flow into numerous arteries of the large and small intestines, and the rectum.

Mesenteric Veins Blood vessels of the digestive system. The superior mesenteric vein drains the small intestine, and the inferior mesenteric collects blood from the colon and rectum. Both deliver their blood to the splenic vein.

Mesentery A tissue that suspends the digestive glands within the abdominal cavity. The mesentery connects to the outer layer of the gastrointestinal tract.

Metabolism The sum of all of the chemical reactions in a cell, tissue, organ, or organism. In nutritional terms, it frequently applies to the processing of the energy nutrients and generation of energy.

Microvilli Small outgrowths covering the intestinal villi. They increase the surface area of the villi, aiding in nutrient uptake by capillaries.

Micturition The process in which urine is released from the bladder; urination.

Midsagittal plane An imaginary line that passes through the skull and spinal cord, dividing the body into equal halves.

Mineralocorticoids A class of hormones produced by the adrenal cortex that regulate mineral metabolism.

Minor Calyx A cup-like receptacle attached to each renal pyramid in the kidney.

Mitochondria Located in the cell's cytoplasm, these are organelles where cell respiration takes place and energy is produced.

Mitosis A process of cell division resulting in daughter cells containing the same number of chromosomes as the parent cell.

Monocyte A type of white blood cell. They become macrophages, large cells that engage in phagocytosis.

Monozygotic Refers to twins arising from one ovum.

Morula A compacted group of embryonic cells at a level of development between the zygote and blastocyst stages.

Motilin A gastrointestinal hormone that stimulates intestinal muscle contractions to clean undigested materials from the small intestine.

Mucociliary Pertaining to mucus and to the cilia of the epithelial cells in the respiratory system.

Mucosa A mucous membrane that lines a body cavity.

Multiple Marker Test Testing to screen for various biomarkers of disease. *See also* Biomarker.

Muscle Fiber A muscle unit made up of many muscle cells that have fused together and received the signal to contract from a single nerve.

Muscle Spindle Related to the stretch reflex, this receptor responds to the muscle's passive stretch and contraction. The muscle spindles are parallel with the muscle fibers.

Myelin Sheath A substance composed of fatty material that covers most axons and dendrites in the central and peripheral nervous systems in order to electronically insulate neurons from one another.

Myofibril The contractile unit within a muscle fiber that is made up of a series of contractile units called sarcomeres. Each muscle fiber contains many myofibrils, all of which contract when the muscle fiber receives a signal from a nerve.

Myoglobin A molecule in the muscle that collects oxygen from the blood and delivers it to mitochondria in the muscle fiber.

Myosin One of the major contractile proteins making up a muscle fiber. Myosin proteins form chains that pull on actin filaments, causing the muscle fiber to contract.

Nephron The filtering unit of the kidney.

Nerve A system of neurons with blood vessels and other connective tissue.

Nerve Fiber The neuron including the axon and the surrounding cells. These fibers branch out at the neuron's ending, which is known as arborization.

Nerve Plexus A combination of neurons from various sections of the spinal cord that serve specific areas of the body.

Nerve Tracts A neuron group that performs a common function in the central nervous system. This grouping can be ascending (sensory) or descending (motor).

Neurohormone A chemical messenger released by the hypothalamus that signals the pituitary gland to release or inhibit release of its hormones.

Neurolemma Essential to the regeneration of damaged neurons in the peripheral nervous system, this is a sheath surrounding peripheral axons and dendrites and is formed by cytoplasm and the nuclei of Schwann cells.

Neuron A nerve cell that consists of a cell body, in addition to an axon and dendrites.

Neurosecretory Cells Specialized nerve cells that transmit chemical impulses, release hormones, and serve as a link between the endocrine and nervous systems.

Neurotransmitter Chemical substances that are emitted through nerve endings to help transmit messages. In the human body, there are about 80 different neurotransmitters.

Neutrophil The most common type of granulocyte. Neutrophils are a main bodily defense mechanism against infection, and are particularly suited to engulfing and destroying bacteria, although they can also combat other small invading organisms and materials.

Node of Ranvier The cell region located on or between the Schwann cells.

Noradrenalin A type of neurotransmitter that transports neurons throughout the various regions in the brain and spinal cord, in addition to increasing the reaction excitability in the CNS and the sympathetic neurons in the spinal cord.

Norepinephrine A hormone that causes blood pressure to rise in stressful situations.

Normoblast The cells of the bone marrow that are responsible for the formation of the red blood cells.

Nucleus The cell's largest organelle that contains chromosomes and hereditary material.

Occipital Lobes The most posterior part of the brain, containing the visual areas.

Oligodendrocytes A type of neuroglia that forms the neuron's myelin sheath.

Oocytes Ova that have not yet matured in the ovary; they arise from primordial oogonia that develop in the fetus.

Oogenesis The formation and development of an egg in the ovary.

Oogonia Cells that arise from primordial germ cells and differentiate into oocytes in the ovary.

Opposable Thumb In primates including humans, the ability to use the thumb to touch each finger.

Opsonization The modification of a bacterium so that it is more easily recognized by the immune system, resulting in an increase in phagocytosis by macrophages.

Organelles Primary components in a cell, including the nucleus, chromosomes, cytoplasm, and mitochondria.

Organic Molecules Molecules that contain carbon-carbon or carbon-hydrogen bonds.

Origin The end of the muscle closest to the body.

Osmoreceptors Neurons that sense fluid concentrations and send a message to the hypothalamus.

Osmosis A process that seeks to equalize the water-to-solute ratio on each side of a water-permeable membrane.

Osteology The study of bones, from the Greek word osteon, meaning "bone" and the suffix -ology, meaning "study of."

Ovarian Vein One of a pair of veins serving the female reproductive system.

Oxaloacetic acid An acid formed by oxidation of maleic acid, as in metabolism of fats and carbohydrates in the Krebs cycle.

Oxidation-reduction Reaction A reaction in which there is transfer of electrons from an electron donor (the reducing agent) to an electron acceptor (the oxidizing agent). Also called the redox reaction. In the electron transport system, this reaction results in molecules alternately losing and gaining an electron.

Oxidative Phosphorylation The process of combining electrons with oxygen to create water. This process also produces energy for the cell when enough oxygen is present.

Oxidization Add oxygen to or combine with oxygen, usually in chemical processes.

Oxygen Dissociation Curve A graph that shows the percent saturation of hemoglobin at various partial pressures of oxygen. The curve shifts to the right (the Bohr effect) when less than a normal amount of oxygen is taken up by the blood and shifts to the left (the Haldane effect) when more than a normal amount is taken up.

Pacemaker *See* SA Node.

Palmar Indicates the palms of the hands.

Pancreatic Polypeptide Hormone secreted by the F cells of the endocrine pancreas that inhibits gallbladder contraction and halts enzyme secretion by exocrine cells in the pancreas.

Pandemic An epidemic that occurs over a large geographic area, sometimes throughout the world.

Papillary Duct A tube that drains urine from collecting ducts in the nephron and empties it into the minor calyx.

Paracrine The action of a hormone on neighboring cells.

Parasympathetic Nervous System Also known as the vagal system. It is one of two major divisions of the autonomic nervous system. It functions to inhibit the pacemaker and lower the heart rate. *See also* Sympathetic Nervous System.

Parathyroid Hormone (PTH) A hormone secreted by the parathyroid gland that helps maintain calcium and phosphorous levels in the body. PTH controls the release of calcium from bone, the absorption of calcium in the intestine, and the excretion of calcium in the urine. Also called parathormone.

Partial Pressure Within the circulatory system, it is a term used to describe the relative oxygen concentration in tissues. For example, hemoglobin has a differential ability to bind oxygen: It picks up oxygen when the partial pressure in surrounding tissues is high, as it is in the lungs, and drops off oxygen when the partial pressure in the surrounding tissues is low, as it is in the tissues.

Pathogens Disease-producing agents such as virus, bacterium, or other microorganisms.

Peptides A chemical that helps to join amino acids in a protein molecule.

Pericardium The two-layered membranous sac around the heart.

Perimysium Connective tissue that surrounds the bundle of muscle fibers making up a fascicle.

Peripheral Nervous System Division of the nervous system that consists of the spinal and cranial nerves.

Peristaltic Action A rhythmic contraction of the muscles of the gastrointestinal tract, most notably in the small intestine, that is responsible for moving nutrients and undigested material through the lumen towards the anus.

Peroneal Veins Also known as fibular veins. They drain the lower leg and ankle, and deliver the blood to the posterior tibial vein.

pH The acidity of a solution. It is formally the measure of the hydrogen ion concentration of a solution.

Phagocytic Cell A type of cell that engulfs external particles, food, or organisms into its cytoplasm; the enclosed material may then be destroyed by digestive enzymes.

Phagocytosis The process of engulfing and destroying bacteria and other antigens.

Pharynx The rear area of the oral cavity. This area connects the respiratory and digestive systems of the body.

Phosphate A chemical related to energy usage and transmission of genetic information in the cell.

Phospholipids A class of organic molecules that resemble triglycerides but have one fatty acid chain replaced by a phosphate group.

Pia Mater The meninges' innermost layer, made of thin connective tissue located on the surface of the brain and spinal cord.

Pituitary Gland An endocrine gland at the base of the brain that sends out growth hormones.

Plantar Indicates the soles of the feet.

Plasma The liquid portion of blood in which red and white blood cells, platelets, and other blood contents float.

Plasminogen A beta globulin that participates in blood clotting.

Plasticity The reorganization of the nervous system following an injury or a tissue-damaging disease.

Platelets Also known as thrombocytes. They are round or oblong disks in the blood that participate in blood clotting.

Pleura A membrane that envelops the lung and attaches the lung to the thorax. There are two pleurae, right and left, that are entirely distinct from each other, and each pleura is made of two layers. The parietal pleura lines the chest cage walls and covers the upper surface of the diaphragm, and the visceral pleura tightly covers the exterior of the lungs. The two layers are actually one continuous sheet of tissue that lines the chest wall and doubles back to cover the lungs. The pleura is moistened with a thin, serous secretion that helps the lungs to expand and contract in the chest.

Polarization A chemically charged state when the neuron's membrane has a positive charge outside and a negative charge inside.

Polyploid Cells In humans, each cell has two copies of each chromosome, one maternal and one paternal. Polyploid indicates more than two chromosomes in a cell.

Polyspermy The entrance of several sperm into an ovum.

Polyunsaturated Fatty Acids Components of dietary fats that contain at least two double bonds.

Pons The parts of the brain that are anterior and superior to the medulla. The pons regulate respiration.

Popliteal Artery A blood vessel that arises from the femoral artery and traverses the knee before dividing into the posterior and anterior tibial arteries.

Popliteal Vein A blood vessel that collects blood from the anterior and posterior tibial veins, and empties into the femoral vein.

Porphyrin A complex, nitrogen-containing compound that makes up the various pigments found in living tissues. Iron-containing porphyrins are called hemes.

Portal Circulation *See* In-series Blood Circulation.

Portal Vein A blood vessel that arises from the splenic vein and superior mesenteric vein. It empties into the liver.

Positron Emission Tomography (PET) Scan A type of brain imaging technique that shows the brain in action. In order to obtain this image, a radioactive substance (such as glucose) is injected into the brain and then followed as it moves throughout the brain.

Posterior Indicates the back of a person or mammal.

Posterior Pituitary Lobe of the pituitary gland that is an extension of the nervous system.

Postganglionic Neuron A neuron located in the autonomic nervous system that extends from a ganglion to the visceral effector.

Postsynaptic Any impulse event following transmission at the synapse.

Preganglionic Neuron A neuron located in the autonomic nervous system that extends from the central nervous system to a ganglion and then synapses with a postganglionic neuron.

Pregnenolone A steroid hormone precursor produced from cholesterol.

Preprohormone/Prohormone An inactive sequence of amino acids from which an active hormone is released.

Progesterone Steroid hormone produced in the adrenal gland, placenta, and corpus luteum that influences sexual development and reproduction.

Progestin Female hormone produced by the ovaries that influences sexual development and pregnancy.

Proglucagon Precursor molecule from which the hormone glucagon is produced.

Proinsulin The inactive precursor molecule from which insulin is formed.

Projection A sensory occurrence when the sensation is felt in the receptor area.

Prolactin A protein hormone secreted by the anterior pituitary that stimulates mammary gland development and milk production.

Prostaglandin Fatty acid derivatives that act much like hormones to influence a number of physiological processes throughout the body.

Prostate The gland surrounding the top of the urethra in men that contributes nutrients to the seminal fluid.

Protease A class of enzyme that is involved in the breakdown of proteins into amino acids.

Protein Complex chemical compounds that are essential to life.

Prothrombin A beta globulin that participates in blood clotting.

Protozoa Single-celled, eukaryotic organisms, including many parasites.

Proximal Indicates direction closer to the torso.

Proximal Convoluted Tubule (PCT) Tiny tubes in the nephrons of the kidneys through which glomerular filtrate passes and substances necessary to the body (i.e., water, sodium, and calcium) are reabsorbed into the bloodstream.

Pulmonary Artery The blood vessel that originates at the right ventricle, then splits into two branches. The left and right pulmonary arteries lead to the left and right lung, respectively.

Pulmonary Circulation The transit of blood from the heart to the lungs and back to the heart. Blood picks up oxygen and drops off carbon dioxide in this circulatory route.

Pulmonary Semilunar Valve The three-cusped heart valve located between the right ventricle and pulmonary artery.

Pulmonary Veins Four blood vessels that flow from the lungs to the left atrium.

Purkinje Fibers A mesh of modified muscle fibers located in the base of the ventricle wall. The fibers receive the electrical impulse from the bundle of His and deliver it to the ventricle, which then contracts.

Pyruvate The end product of glycolysis.

Radial Artery A blood vessel in each lower arm that receives blood from the brachial artery and delivers it to numerous arteries of the forearm, wrist, and hand.

Radial Vein A blood vessel in each arm that collects blood from veins in the hand. It eventually merges with the ulnar vein into the brachial vein.

Receptors Proteins on the surface of cells or within cells that bind to particular hormones.

Rectal Vein Blood vessels in the digestive system that drain parts of the rectum. The inferior rectal vein joins the internal pudendal vein, which flows into the internal iliac vein, while the middle rectal vein connects directly to the internal iliac vein. The superior rectal vein flows directly into the inferior mesenteric vein.

Red Blood Cells Also known as erythrocytes. These are the cells in the blood that are responsible for gathering and delivering oxygen and nutrients to the body tissues, and for disposing of the tissue's waste products.

Reflex An automatic or involuntary response to a stimulus.

Renal Artery A pair of blood vessels that arise from the abdominal aorta. Each feeds a kidney and adrenal gland, and the ureter.

Renal Fascia The outermost layer of the kidney, composed of connective tissue that holds the kidney to the abdominal wall.

Renal Pelvis A funnel-shaped cavity that collects urine and sends it into the ureter.

Renal Pyramids Cone-shaped receptacles inside the medulla of the kidney.

Renal Veins A pair of blood vessels that drain the two kidneys. They empty into the inferior vena cava.

Renin An enzyme secreted by the kidneys that leads to the production of the hormone aldosterone.

Repolarization A chemically charged state following a neuron's depolarization, when the membrane has a positive charge outside and a negative charge inside due to the outflow of potassium ions.

Respiration In the respiratory system, the movement of respiratory gases, such as oxygen and carbon dioxide, into and out of the lungs.

Reticulocyte Immature red blood cells; these are usually found in the bone marrow.

Rh Factor An antigen that is found on the surface of blood cells; it is an independent factor of the ABO group.

Rotation Involves turning a body part on an axis.

Saggital Plane An imaginary vertical line that divides the body into right and left segments.

SA Node Also known as the sinoatrial node, or pacemaker. This is a group of small and weakly contractile modified muscle cells that spontaneously deliver the electrical pulses that trigger the heart's contraction.

Sarcolemma The cell membrane of a muscle fiber.

Sarcomere An individual contractile unit within the myofibril that contains actin filaments attached to either end. Myosin chains pull the actin filaments closer together, making the sarcomere grow shorter.

Sarcoplasmic Reticulum A network of tubules that runs throughout the muscle fiber. The sarcoplasmic reticulum stores calcium when the fiber is not contracted and releases calcium when the fiber receives a signal to contract.

Schwann Cells Located in the peripheral nervous system, these cells form the myelin sheath and neurolemma of the peripheral axons and dendrites.

Semilunar Valves Valves, shaped like half-moons, that ensure blood movement in only one direction. They are found in the heart and in large blood vessels.

Seminiferous Tubules Tubes in the testes in which sperm are produced.

Semipermeable (or Selectively Permeable) Membrane A membrane that allows certain molecules to pass through while restricting others.

Sensory Nerves A type of afferent nerve coming in at the back of the spinal cord; also called posterior nerves.

Sensory Neurons Also known as afferent neurons, they carry impulses and messages to the spinal cord and brain.

Septum A partition, dividing wall, or membrane that separates bodily spaces or masses of tissue. In the respiratory system, septum most often refers to the cartilage separating the two nostrils.

Serosa The outer layer of the bladder wall.

Serotonin A neurotransmitter present throughout the central nervous system.

Sertoli Cells Cells in the testes in which sperm is produced.

Sesamoid Bone Short bones embedded within a tendon or joint capsule.

Sex-linked Inherited Characteristics Traits, such as color-blindness, that are linked to genes on the sex chromosomes, especially the X chromosome.

Sickle Cell Anemia A serious autosomal recessive disease characterized by abnormal red blood cells.

Sigmoidal Artery A blood vessel that arises from the inferior mesenteric artery and supplies blood to the lower abdominal region.

Skeletal Muscle Muscles that are attached to the skeleton and allow the body to move. This is also called voluntary muscle because these are the muscles that move voluntarily.

Slow-twitch Muscles A type of muscle fiber that is able to contract very quickly. These are predominantly found in muscles that must contract repeatedly but without much strength.

Smooth Muscle Also known as an involuntary muscle. It is a type of muscle that is controlled by the autonomic nervous system, rather than by willful command, as is the striated muscle.

Sodium/Potassium Pump A form of active transport that regulates the amount of sodium and potassium in and around the cells.

Somatic Neuron A type of sensory neuron located in the skeletal muscle and joints.

Somatostatin A hormone produced by the endocrine pancreas and hypo-thalamus that regulates insulin and glucagon release, and inhibits growth hormone release from the pituitary gland.

Somatotroph A cell in the anterior pituitary gland that secretes growth hormone.

Spermatogonia Primordial sperm cells that develop in the male fetus.

Sphincter A skeletal muscle that forms a circular band and that usually controls the size of an opening, such as the mouth or the entrance to the stomach. The muscle contracts to close the opening or relaxes to open it.

Spinal Nerves The spine's 31 pairs of nerves located in the peripheral nervous system.

Spinal Reflex An automatic or involuntary reflex related to the spinal cord and in which the brain is not directly involved.

Splenic Artery Blood vessel that arises from the celiac trunk and branches into numerous arteries that feed the stomach and peritoneum, pancreas, and spleen.

Splenic Vein A large blood vessel that collects blood from the spleen. It joins the superior mesenteric vein to create the portal vein.

Stem Cells Undifferentiated cells. They have the genetic potential to mature into specific cell types. Some stems are only able to become one type of cell, while others have the ability to become any number of different cells.

Stimulus Any sort of change in a living organism that causes a response or affects a sensory receptor.

Stretch Reflex A reflex from the spinal cord in which a muscle will respond to a stretch by contracting.

Striated Muscle Also known as voluntary muscle. A person can consciously control the action of striated muscle.

Subclavian Arteries Blood vessels that supply the arms, much of the upper body, and the spinal cord. The right subclavian artery branches from the brachiocephalic artery, while the left divides off of the aortic arch. Numerous arteries arise from each.

Subclavian Veins Primary blood vessels draining the arms. They collect blood from the axillary vein and later merge with the internal jugular vein to produce the brachiocephalic vein.

Substance P Neuropeptide found in the gut and brain that stimulates smooth muscle contraction and epithelial cell growth and that plays a role in both the pain and pleasure responses.

Sulci Grooves between the gyri of the cerebellum.

Superior Direction given to a body part that indicates toward the head.

Surfactant A substance that acts on the surface of objects. In the respiratory system, surfactants are secreted by pneumocyte cells into the alveoli and respiratory air passages, helping to make pulmonary tissue elastic in nature.

Sympathetic Nervous System One of two major divisions of the autonomic nervous system. It functions to stimulate the pacemaker and boost the heart rate. *See also* Parasympathetic Nervous System.

Symphysis A disk of cartilage where two bones meet fiber that attaches a muscle to a bone.

Synapse The junction between two neurons where the axon passes on information to the dendrite. This area is often called a relay because it is here where the information is relayed to the next neuron.

Synaptic Gap or Cleft The actual area (which is approximately 10–50 nanometers in width) between the axon and dendrite where the neurons communicate with each other.

Synarthroses Nonmoveable joints.

Synergist A muscle that works in conjunction with an antagonistic pair to control the movement of a joint. The synergist usually runs beside a joint or diagonally across a joint.

Synovial Fluid The clear fluid that is normally present in joint cavities.

Systemic Circulation The transit of blood from the heart to the body (except the lungs) and back to the heart. *See also* Coronary Circulation and Pulmonary Circulation.

T Cells Also known as T lymphocytes. They are one of two main types of lymphocyte, and participate in the body's immune response.

T Tubule Tubules that run through muscle fibers carrying the signal to contract. The signal passes from the T tubule to the sarcoplasmic reticulum, which releases calcium and causes the contraction to take place.

Target Cells Cells that are responsive to a particular hormone.

Tendon A band of connective tissue that connects the muscle to the bone.

Terminal Arterioles Arterioles that feed capillaries.

Testosterone A hormone that produces male characteristics including large muscles.

Tetanus Contraction A sustained contraction as a result of many independent signals from a nerve.

Thalamus The portion of the brain located superior to the hypothalamus that controls the elements of subconscious sensation.

Threshold Level This value in a nerve fiber depends on the composition of the cellular fluid and the number of impulses recently received and conducted. When this level is reached in the nerve fiber's axon, a reaction results.

Thrombocytes *See* Platelets

Thromboplastin A substance released by damaged tissue and platelets. With calcium, it promotes the formation of blood clots.

Thyroid-stimulating Hormone (TSH) Hormone produced by the pituitary gland that stimulates the thyroid gland to secrete its hormones, thyroxine (T4) and triiodothyronine (T3). Also called thyrotropin.

Thyrotroph Cell in the anterior pituitary gland that secretes thyroid-stimulating hormone.

Thyrotropin-releasing Hormone (TRH) Hypothalamic neurohormone that triggers the release of thyroid-stimulating hormone (TSH) and prolactin (PRL) from the pituitary gland.

Thyroxine (T4) Thyroid hormone that influences metabolism and growth.

Tibial Arteries Blood vessels of the lower leg. The posterior and anterior tibial arteries arise from the popliteal artery and supply blood to arteries feeding the lower leg, ankle, and foot.

Tibial Veins Blood vessels of the lower leg. The anterior and posterior tibial veins drain the leg, then join together to form the popliteal vein.

Tonsils The name given to the lymphatic tissue found at the back of the oral cavity.

Toxoid The toxin produced by a bacterium that has been detoxified, but still retains its antigen characteristics. Toxoids are useful in the generation of immunizations.

Trabeculae Beams that act as strengthening girders of cancellous bone.

Trabecular Bone The porous, spongy bone that lines the bone marrow cavity and is surrounded by cortical bone.

Transverse Plane An imaginary line passing at right angles to both the front and midsection; a cross section.

Trigone A triangular-shaped region located in the bladder floor.

Triiodothyronine (T3) The more potent of the two thyroid hormones.

Tropomyosin A protein that forms long filaments wrapping around actin within the muscle fiber.

Troponin A protein that is associated with actin and tropomyosin within the muscle fiber.

Tunica Adventitia Fibrous connective tissue forming the outer of the three layers comprising arteries, arterioles, veins, and venules. *See also* Tunica Intima and Tunica Media.

Tunica Intima Also known as endothelium. It forms the innermost of the three layers comprising arteries, arterioles, veins, and venules. Capillaries are composed of only a single layer of endothelial cells. *See also* Tunica Adventitia and Tunica Media.

Tunica Media Muscular and elastic tissue forming the middle of the three layers comprising arteries, arterioles, veins, and venules. *See also* Tunica Adventitia and Tunica Intima.

Type A Blood Blood containing a certain antigen called "A." Due to potential antigen reactions, a person with type A blood can receive blood donations of type A and type O, but not type B or type AB.

Type AB Blood Blood containing anti-lymphocytes. *See also* T Cells.

Type B Blood Blood containing a certain antigen called "B." Due to potential antigen reactions, a person with type B blood can receive blood donations of type B and type O, but not type A or type AB.

Type O Blood Blood containing neither of the antigens called "A" and "B." Due to potential antigen reactions, a person with type O blood can receive blood donations of type O, but not type A, type B, or type AB.

Tyrosine An amino acid component of protein.

Ultrasound Scan An imaging method using high-frequency sound waves to form images inside the body. Also called ultrasonography.

Urea Waste produced by the breakdown of proteins.

Ureter A long tube that delivers urine from the kidney to the bladder.

Ureteral Orifices Two holes where the ureters pierce the bladder.

Urethra A muscular tube that connects the bladder with the exterior of the body.

Uric Acid Waste produced by the breakdown of nucleic acids (DNA and RNA).

Urochrome Pigment produced by the breakdown of bile that gives urine its yellow or amber color.

Vagus Nerve The 10th of 12 cranial nerves, which originates somewhere in the medulla oblongata in the brainstem and extends down to the abdomen.

Vasoconstrictor Nerves Nerves that signal the veins to constrict.

Vasodilation The relaxation of the muscles surrounding the vascular tissue; this increases the diameter of the vessel and reduces pressure.

Vasopressin A hormone produced by the pituitary gland that increases the permeability of the kidney ducts to return more fluid to the bloodstream. Also called antidiuretic hormone (ADH).

Vena Cava One of two large veins, the superior and inferior venae cavae, bringing blood from the body back to the heart.

Ventral Root The motor root of a spinal nerve that attaches the nerve to the anterior part of the spinal cord.

Ventricle In the human heart, it is one of the heart's two lower chambers.

Vertebral arteries A pair of blood vessels on each side of the neck that arise from the subclavian arteries. They unite at the basilar artery.

Vestibule The opening or entrance to a passage, such as the vestibule of the vagina.

Vestigial A term for nonfunctional remnants of organs.

Virus A nonliving infectious agent that is characterized as having a protein covering and either DNA or RNA as its genetic material; some viruses may also have a lipid covering. Viruses are completely dependent on cells for reproduction.

Visceral Neuron A type of sensory neuron located in the body's internal organs.

Visceral Organs The body's internal organs, such as the heart and lungs, that have nerve fibers and nerve endings that conduct messages to the brain and spinal cord.

White Blood Cells Also known as leukocytes. These are the cells in the blood that function in the body's defense mechanism to detect, attack, and eliminate foreign organisms and materials.

White Matter The nerve tissue located within the central nervous system that contains myelinated axons and interneurons.

Z Band A dense area that separates the sarcomeres. The actin filaments are embedded in the Z band, extending inward into each sarcomere.

Zona Fasciculate The middle layer of the adrenal cortex, in which the glucocorticoids (cortisol) are produced.

Zona Glomerulosa The outermost layer of the adrenal cortex, in which the mineralocorticoids (aldosterone) are produced.

Zona Pellucid The outer covering of an ovum.

Zona Reticularis The innermost layer of the adrenal cortex, in which the gonadocorticoids (sex hormones) are produced.

Zygote A diploid cell resulting from fertilization of an egg by a sperm cell.

Select Bibliography

Aaronson, Philip I., and Jeremy P. T. Ward, with Charles M. Wiener, Steven P. Schulman, and Jaswinder S. Gill. *The Cardiovascular System at a Glance*. Oxford: Blackwell Science Limited, 1999.

Abrahams, Peter, ed. *How the Body Works*, London: Amber Books, 2009.

"Acne." National Institute of Arthritis and Musculoskeletal and Skin Diseases. http://www.niams.nih.gov/Health_Info/Acne/default.asp (accessed June 20, 2010).

Adams, Amy. *The Muscular System*. Westport, CT: Greenwood Publishing, 2004.

"Alcohol-Induced Liver Disease." American Liver Foundation, http://www.liverfoundation.org/abouttheliver/info/alcohol/ (accessed June 20, 2010).

American Academy of Allergy, Asthma and Immunology. http://www.aaaai.org (accessed June 20, 2010).

American Academy of Family Physicians. http://www.familydoctor.org (accessed June 20, 2010).

American Academy of Otolaryngology. http://www.entnet.org (accessed June 20, 2010).

Asimov, Isaac. *The Human Body: Its Structure and Operation*. Rev. ed. New York: Mentor, 1992.

Bainbridge, David. *Making Babies: The Science of Pregnancy*. Cambridge, MA: Harvard University Press, 2001.

"Bariatric Surgery for Severe Obesity." National Institute of Diabetes and Digestive and Kidney Diseases, Weight-Control Information Network. http://win.niddk.nih.gov/publications/gastric.htm (accessed June 20, 2010).

Bastian, Glenn F. *An Illustrated Review of the Urinary System.* New York: HarperCollins College Publishers, 1994.

Berne, Robert M., and Matthew N. Levy. *Cardiovascular Physiology.* 6th ed. St. Louis, MO: C. V. Mosby-Year Book, 1992.

Charlton, C. A. C. *The Urological System.* Harmondsworth, UK: Penguin Books, 1973.

Cornett, Frederick D., and Pauline Gratz. *Modern Human Physiology.* New York: Holt, Rinehart, and Winston, 1982.

"Did You Know . . . Facts about the Human Body." Health News. http://www.healthnews.com (accessed June 20, 2010).

"Drinking Water." Centers for Disease Control and Prevention. http://www.cdc.gov/healthywater/drinking/travel/index.html (accessed June 20, 2010).

"Flu." Centers for Disease Control and Prevention. http://www.flu.gov (accessed June 20, 2010).

"Fun Science Facts." High Tech Science. http://www.hightechscience.org/funfacts.htm (accessed June 20, 2010).

Gilbert, S. F., M. S. Tyler, and R. N. Kozlowski. *Developmental Biology,* 6th ed. Sunderland, MA: Sinauer Associates, 2000.

"Global Water, Sanitation, and Hygiene (WASH)." Centers for Disease Control and Prevention. http://www.cdc.gov/healthywater/global/index.html (accessed June 20, 2010).

Greenspan, Francis S., and David G. Gardner. *Basic and Clinical Endocrinology.* 6th ed. New York: Lange Medical Books/McGraw-Hill, 2001.

"The Heart: An Online Exploration." http://sln.fi.edu/biosci/heart.html (accessed June 20, 2010).

Hess, Dean, and Robert M. Kacmarek. *Essentials of Mechanical Ventilation.* 2nd ed. New York: McGraw-Hill, Health Professions Division, 2002.

Hlastala, Michael P., and Albert J. Berger. *Physiology of Respiration.* 2nd ed. New York: Oxford University Press, 2001.

Hollen, Kathryn. *The Reproductive System.* Westport, CT: Greenwood Publishing, 2004.

Holmes, Oliver. *Human Neurophysiology: A Student Text.* 2nd ed. London: Chapman & Hall Medical, 1993.

"How Does Smoking Affect the Heart and Blood Vessels?" National Heart and Lung Institute. http://www.nhlbi.nih.gov/health/dci/Diseases/smo/smo_how.html (accessed June 20, 2010).

"The Human Body." Teachnology. http://www.teach-nology.com/themes/science/humanb/ (accessed June 20, 2010).

"Interesting Facts about the Human Body." Random Facts. http://facts.randomhistory.com/2009/03/02_human-body.html (accessed June 20, 2010).

Kelly, Evelyn. *The Skeletal System.* Westport, CT: Greenwood Publishing, 2004.

Knight, Bernard. *Discovering the Human Body.* New York: Lippincott & Crowell, 1980.

"LASIK." Food and Drug Administration. http://www.fda.gov/MedicalDevices/ProductsandMedicalProcedures/SurgeryandLifeSupport/LASIK/default.htm (accessed June 20, 2010).

Lyman, Dale. *Anatomy DeMystified.* New York: McGraw-Hill, 2004.

"Massage Therapy: An Introduction." National Center for Complementary and Alternative Medicine. http://nccam.nih.gov/health/massage/ (accessed June 20, 2010).

McDowell, Julie. *The Nervous System and Sensory Organs.* Westport, CT: Greenwood Publishing, 2004.

McDowell, Julie, and Michael Windelspecht. *The Lymphatic System.* Westport, CT: Greenwood Publishing, 2004.

"Medical References." University of Maryland Medical Center. http://www.umm.edu/medref/ (accessed June 20, 2010).

Mertz, Leslie. *The Circulatory System.* Westport, CT: Greenwood Publishing, 2004.

"National Cholesterol Education Program." National Heart Lung and Blood Institute. http://www.nhlbi.nih.gov/chd/ (accessed June 20, 2010).

"National Diabetes Statistics, 2007." National Institute of Diabetes and Digestive and Kidney Diseases. http://diabetes.niddk.nih.gov/dm/pubs/statistics/index.htm#what (accessed June 20, 2010).

National Institute of Allergy and Infectious Diseases. http://www.niaid.nih.gov (accessed June 20, 2010).

Northwestern University Medical School, Department of Neurology. http://www.neurology.northwestern.edu/ (accessed June 20, 2010).

Petechuk, David. *The Respiratory System*. Westport, CT: Greenwood Publishing, 2004.

Phillips, Chandler A., and Jarold S. Petrofsky. *Mechanics of Skeletal and Cardiac Muscle*. Springfield, IL: Thomas, 1983.

Sanders, Tina, and Valerie C. Scanlon. *Essentials of Anatomy and Physiology*. 3rd ed. Philadelphia: F. A. Davis Company, 1999.

Sherwood, Lauralee. *Human Physiology: From Cells to Systems*. 4th ed. Pacific Grove, CA: Brooks/Cole, 2001.

Soloman, Eldra P., Linda R. Berg, Diana W. Martin, et al. *Biology*. 4th ed. Orlando, FL: Harcourt Brace & Company, 1997.

"Spinal Cord Research." Christopher and Dana Reeve Foundation. http://www.christopherreeve.org/site/c.ddJFKRNoFiG/b.4343879/k.D323/Research.htm (accessed June 20, 2010).

"Sports Injuries." National Institute of Arthritis and Musculoskeletal and Skin Diseases. http://www.niams.nih.gov/Health_Info/Sports_Injuries/default.asp (accessed June 20, 2010).

Steele, D. Gentry, and Claude A. Bramblett. *The Anatomy and Biology of the Human Skeleton*. College Station: Texas A&M University Press, 1988.

Takahashi, Takeo. Atlas of the Human Body. New York: HarperCollins Publishers, 1989.

Watson, Stephanie. *The Endocrine System*. Westport, CT: Greenwood Publishing, 2004.

Watson, Stephanie. *The Urinary System.* Westport, CT: Greenwood Publishing, 2004.

"What Is Coronary Disease?" National Heart and Lung Institute. http://www.nhlbi.nih.gov/health/dci/Diseases/Cad/CAD_WhatIs.html (accessed June 20, 2010).

Windelspecht, Michael. *The Digestive System.* Westport, CT: Greenwood Publishing, 2004.

Index